Data Modeling

Elmar Rueckert

Data Modeling

From Physical Processes to Machine Learning

Springer

Elmar Rueckert
Institute of Cyber-Physical-Systems
Technical University of Leoben
Leoben, Austria

ISBN 978-981-95-7591-6 ISBN 978-981-95-7592-3 (eBook)
https://doi.org/10.1007/978-981-95-7592-3

This Springer imprint is published by the registered company Springer Nature Singapore Pte Ltd.
The registered company address is: 152 Beach Road, #21-01/04 Gateway East, Singapore 189721, Singapore

He who loves practice without theory is like the sailor who boards a ship without a rudder and compass and never knows where he may cast.

—*Leonardo da Vinci (1452–1519)*

Die Welt zu verstehen und zu beschreiben ist seit jeher eines unserer höchsten Ziele. Auch im Zeitalter der Digitalisierung hat sich daran nichts geändert, nur die Werkzeuge, die wir nützen.

—*Elmar Rückert (2025)*

Preface

This textbook introduces the fundamental theories, concepts, and methods of data modeling, with a particular focus on linking physical processes to data-driven and machine learning–based predictions. Rather than aiming for exhaustive or highly specialized coverage, the book is designed to support a **single semester course for first-semester undergraduate students**. Its primary goal is to convey the essential principles, highlight dependencies between topics, and demonstrate how the different elements of data modeling interact within a coherent framework.

The book covers core aspects of data modeling, including data collection, organization and storage, information representation, data analysis, and introductory machine learning applications. Emphasis is placed on building intuition and understanding, enabling students to grasp not only individual techniques but also their role within the broader modeling pipeline.

Designed for students with little or no prior background in data science or machine learning, the material introduces key concepts in a clear and accessible manner. Hands-on examples and practical exercises encourage active engagement and illustrate how data-driven decision-making and predictive models emerge from real-world measurements.

The content is aligned with the lecture series of Univ.-Prof. Dr. Elmar Rueckert and is further expanded and contextualized during the course. As such, the textbook serves both as a structured companion to the lectures and as a concise reference for students beginning their studies in data modeling and machine learning.

Electronic Supplementary Materials are hosted in the below GitHub link: https://github.com/mul-cps/DMBook

Organization of This Book

The book is organized into eight chapters: 1. Introduction to Data Modeling 2. Fundamentals of Processes and Data Granularity 3. Fundamentals of Sensors 4. Fundamentals of Data 5. Fundamentals of Information Theory 6. Fundamentals of Data Analysis 7. Fundamentals of Data Organization 8. Machine Learning Examples

Leoben, Austria Elmar Rueckert

Acknowledgments The development of this textbook and the accompanying course would not have been possible without the support of many individuals and institutions. I would like to express my sincere gratitude to the Technical University of Leoben for supporting the development of this course from scratch, and for the trust and patience shown throughout this process.

In particular, I would like to thank my colleagues Prof. Peter Moser and Prof. Thomas Prohaska for their support through the provision of student assistant positions, which played a crucial role in the realization of the practical components of this course.

I am grateful to all my students—both in the classroom and in the computer laboratories—who contributed to the development of the content through their engagement, feedback, and curiosity.

Special thanks go to my student assistants Julia Schmelz, Rino Morina, and Tanja Sukal, whose creativity and dedication were instrumental in the design and implementation of the Jupyter Notebooks for the practical exercises. I also thank all my doctoral students for their valuable support in supervising students during these exercises, in particular Melanie Neubauer, Nikolaus Feith, Björn Ellensohn, Vedant Dave, Linus Nwankwo, Puneeth Gottam, Fotios Lygerakis, and Günther Hutter.

Furthermore, I would like to thank my colleagues Eva Anderson, Katharina Binder, Konrad Bartsch, and Kosmo Obermayer for their continuous support in many different forms throughout the development of this course and textbook.

Finally, I am deeply grateful to my family—my wife Eve, and my three daughters Mia, Paula, and Juna—whose enthusiasm and creativity were a great help in designing the "gummi-bear" experiment featured in the introductory chapter.

The curiosity and desire for understanding shared by all these individuals have been a constant source of motivation and inspiration for the creation of this textbook.

Declarations

Competing Interests The author declares that there are no competing interests relevant to the content of this book.

Ethics Approval This book does not report on studies involving human participants or animals. Therefore, no ethics approval or informed consent was required.

Use of AI Tools Artificial intelligence—based tools were used to support grammar checking and language refinement during the preparation of this manuscript. The author reviewed and verified all content to ensure accuracy and integrity.

Contents

Part I
Fundamentals of Data Modeling

This part introduces the fundamental principles of data modeling through historical context, experimentation, and predictive analysis.

Chapter 1
Introduction to Data Modeling

Since ancient times, humans have modeled natural phenomena to better understand the world. This chapter begins with a historical reflection on how Aristotle and later Galileo Galilei shaped early concepts of motion and free fall [27].

We revisit Galileo's inclined plane experiments using a playful, modern adaptation: the "Gummi Bears in Free Flight" experiment. This hands-on activity invites students to investigate whether object mass influences falling time—an idea Galileo famously challenged. Using simple measurement tools and spreadsheet analysis, students gather time and mass data, derive physical equations, and build a predictive model based on linear regression.

The chapter introduces key concepts in data modeling, including model parameter estimation, dataset partitioning (training/test), model generalization, and error analysis. In doing so, it offers a concrete illustration of the scientific and computational thinking behind machine learning.

This introductory chapter lays the foundation for understanding data modeling as a structured process of abstraction, analysis, and interpretation—skills essential for subsequent chapters and future coursework in data science and machine learning.

1.1 From Aristotle to Galileo: The Origins of Data Modeling

Since the dawn of humanity, we have sought to explain natural phenomena through models. Early civilizations, such as the Maya with their sun calendars (around 1000 BCE), created intricate systems to predict astronomical events. Ancient philosophers, like Aristotle in the 4th century BCE, attempted to describe the physical world with logical laws and observations [1, 22].

Aristotle proposed that objects fall to the ground because that is their natural state. He also asserted that heavier objects of the same material fall faster than lighter ones. This idea was accepted for nearly two millennia, largely because it fit

E. Rueckert, *Data Modeling*, https://doi.org/10.1007/978-981-95-7592-3_1

Fig. 1.1 AI generated sketch of the two scientists Aristotle and Galileo Galilei, both studied objects at free fall but at different times

within Aristotle's comprehensive worldview and lacked any rigorous experimental challenge.

It wasn't until the 16th century that Galileo Galilei questioned these long-held beliefs. Through experimentation, Galileo suspected that all objects, regardless of their mass, fall at the same rate in the absence of air resistance. However, due to the limitations of his time—when accurate time-keeping devices were not yet available—he conducted rolling experiments on inclined planes to slow down the motion, allowing for more precise measurements [5, 8]. He then generalized his results to free fall.

Galileo's experiment on free fall is a powerful example of the processes involved in data modeling. In this chapter, we will recreate his experiment using modern tools to gather, analyze, and model data (Fig. 1.1).

To reinforce these concepts, we will extend Galileo's experiment through practical exercises. These activities strengthen understanding of the processes involved in data modeling, which are covered in detail in the subsequent chapters of this book. A comprehensive description of the learning objectives is provided in Appendix B.

1.2 The Gummi Bears in Free Flight Experiment

Data modeling typically starts with a research question or hypothesis. In our case, we want to proof or disproof that Gummi Bears bound together fall faster compared to a single Gummi Bear.

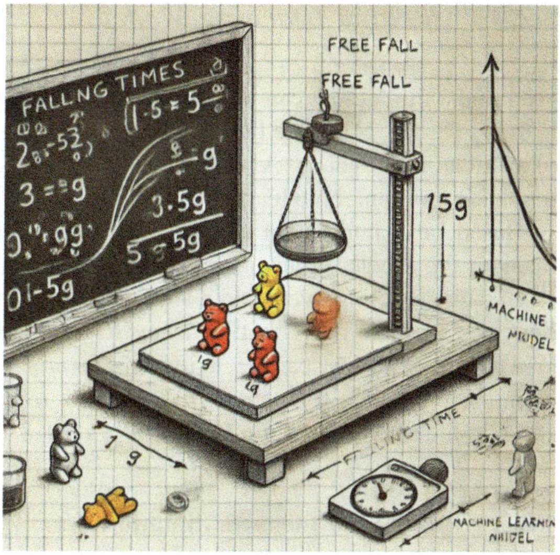

Fig. 1.2 AI generated sketch an experiment studying the effect of object masses on the falling times at the practical example of Gummi Bears in free flight

1.2.1 Experimental Setup

Our setup involves placing individual or grouped Gummi Bears on a platform at a known height. We assume both the height of the platform and the weight of the Gummi Bears can be precisely measured. Additionally, we assume that we can preceisely measure the falling times of the courageous volunteers.

1.2.2 Assumptions

Let's start. We know, from text books [14, 29] or previous experiments, the time-speed law of uniformly accelerated motion $v = at$.[1]

The distance traveled or in our case the height the Gummi Bears fall down can be computed by $h = 1/2at^2$. We are measuring the time t and know the height h. Therefore, we compute the accelerations $a_i = 2h/t_i^2$ for an experiment denoted by the subscript i (Fig. 1.2).

Our hypothesis posits that the acceleration may depend on the object's mass, expressed as $a_i(m_i) = h/t_i^2$.

To proof that we will measure the falling times and will 'store' the observations in a table on the black board, alongside with the previously measured weights of the Gummi Bears.

[1] Note that we do not draw any dots between the variables, this dot would be simply some redundant information and scientists cannot stand redundancies (or model them well).

1.2.3 Machine Learning Model

We build a model that can be learned from data, which is the definition of machine learning [3, 13, 24]. Note that machine learning as a research discipline is a branch of artificial intelligence (AI) focusing on the design and development of algorithms that allow computers to learn from data.

Based on a number of $i = 1 \ldots N$ experiments, we can create a figure that displays the accelerations on the y-axis and the measured weights on the x-axis.

In data modeling, we explain such observations with rules or analytic models. In our case, a simple line equation seems to be an adequate model candidate.[2]

The line equation can be written as $a_i(m_i) = h/(c_1 + c_2 m_i)^2$. Note that this equation is derived form $a_i = h/t_i^2$, where $t_i = c_1 + c_2 m_i$. The line equation is equal to the often used definition of $y = kx + d$ in high school.

1.2.3.1 Model Parameter Estimation

Given our data representation in the form of the table on the blackboard, we can now compute the two open model parameters c_1 and c_2. We can use tools like spreadsheets for that, or we can do that by applying some mathematical concepts.

Mathematically, the solution is given by $\mathbf{c} = (\mathbf{X}^T\mathbf{X})^{-1}\mathbf{X}^T\mathbf{t}$, with $\mathbf{t} = \mathbf{X}\mathbf{c}$, where:

$$\mathbf{t} = \begin{pmatrix} t_1 \\ t_2 \\ \vdots \\ t_N \end{pmatrix}, \quad \mathbf{X} = \begin{pmatrix} 1 & m_1 \\ 1 & m_2 \\ \vdots & \vdots \\ 1 & m_N \end{pmatrix}, \quad \mathbf{c} = \begin{pmatrix} c_1 \\ c_2 \end{pmatrix}.$$

This result is known as the solution for a *linear regression problem*. The term $(\mathbf{X}^T\mathbf{X})^{-1}\mathbf{X}^T$ represents the *Pseudo-Inverse* of the matrix $\mathbf{X}$. A complete derivation can be found in Appendix C.

Example 1.1 For objects with the weights 1 g, 5 g, and 10 g, we recorded falling times in four experiments The measurements are given in Table 1.1.

Using an online matrix calculator, the parameters are found to be

$$\mathbf{c} = \begin{pmatrix} 0.43 \\ 0.003 \end{pmatrix},$$

where the falling times can be reconstructed using $t_i = 0.43 + 0.003 m_i$. Note that the mean squared error of the reconstruction is 6.2 ms. We will analyze this results in detail in Sect. 1.2.6.

---

[2] Note that the model complexity should always be always adapted to the problem complexity. We will discuss that issue in detail in the machine learning chapter of this book.

Table 1.1 Falling times recorded in a free fall experiment for Gummi Bears of different masses

Mass	t_1 [s]	t_2 [s]	t_3 [s]	t_4 [s]
1 g	0.44	0.46	0.43	0.41
5 g	0.39	0.45	0.47	0.44
10 g	0.48	0.46	0.43	0.46

1.2.4 Data Organization

Now that we have an analytic model, it can be used to generate artificial observations or 'predictions'. For any new mass m_i, we can predict the expected acceleration a_i.

To distinguish between the original and new data, we define the **training dataset** as

$$\mathcal{D}_{train} = \{m_i, t_i\}^{i=1...N},$$

and the **test dataset** as

$$\mathcal{D}_{test} = \{m_k, t_k\}^{k=1...K}.$$

The test dataset will be used to analyze the generalization ability. *It is essential that the test data samples were not used for training. This holds for any machine learning application.*

Note that typically a third dataset, namely the **validation dataset**, is created. This dataset is used to optimize the hyper-parameters of a machine learning algorithms. Such hyper-parameters may be learning rates, the number of neurons or layers in neural networks and many more. In our example, there are no hyper-parameters, the line equation model has only two open parameters, which are learned from the data.

1.2.5 Model Generalization

To test our model, using the **test data**, we perform $K = 2$ additional experiments with the masses of 2 g and 20 g. The model is then used to predict the accelerations for these new samples.

The test assesses both **interpolation** (predicting within the range of training data) and **extrapolation** (predicting outside the training range). In machine learning, interpolation generally works well if the test data is within the distribution of training samples, while extrapolation is challenging.

1.2.6 Model Result Interpretation

The machine learning model contains two parameters, c_1 and c_2. The dependency of acceleration a_i on mass m_i is captured through $c_2 m_i$. Interestingly, our experiments show that c_2 is close to zero, **confirming that the falling time and acceleration of objects are independent of mass**, aligning with Galileo's principle.

We can further conclude that the gravity constant with $9.81\,\text{m/s}^2$ can be reconstructed from our model using

$$\begin{aligned} a_i(m_i) &= 2h/t_i^2 \,, \\ &= 2h/(c_1 + c_2 m_i)^2 \ = 2h/0.43^2 \\ &= 10.81 \text{m/s}^2 \,, \end{aligned}$$

where $c_2 = 0$ and the height $h = 1$.

1.3 Process Steps in Data Modeling

The *Gummi Bears in Free Flight Experiment*, demonstrated that by designing a practical experiment, 'storing' measurements in a table, modeling the dependencies of accelerations and weights as a linear system (the machine learning model), by organizing the data into training and test datasets, we were able to proof that Galileo Galilei was right—objects all fall in the same time to the ground, regardless of their mass.

The experiment also shows how these steps are interlinked, forming a comprehensive data modeling system. Figure 1.3 provides an overview of this process, which will serve as a guiding framework throughout this book.

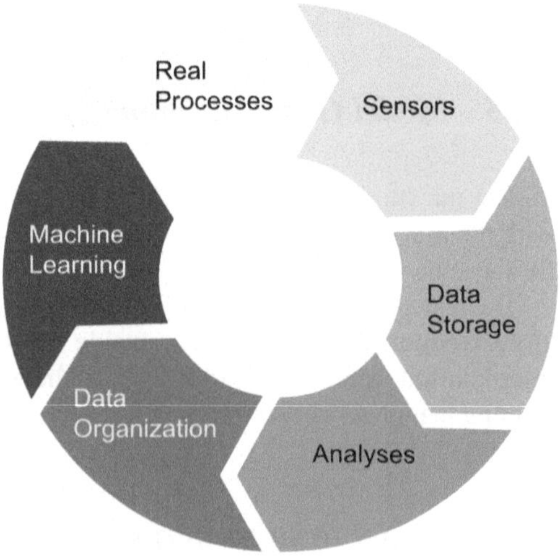

Fig. 1.3 Six processes or steps are involved in data modeling. This book discusses all of the components in detail in individual chapters

1.4 Practical Exercises: Data Modeling of Objects in Free Flight

In these exercises, students will explore the complete process chain, from storing and processing data to finally generating predictions with machine learning models. The limitations of simple linear models will be identified, and solution strategies based on non-linear models will be explored.

1.4.1 Exercise 1: Data Exploration and Model Requirement Identification

Objective: Learn how to inspect data with respect to the model complexity.

Given is a tabular description of measurement of a physical process. Implement the following tasks:

- Load the provided dataset containing measurements of falling times (t_i) at different heights (h_i).
- Display the measurements in a tabular form using the pre-implemented code.
- Visualize the measurements using a scatter plot.

Answer the following questions:

- What can you observe from a visual inspection of the height and time variables in the data presented in tabular form?
- What can you observe from a visual inspection of the scatter plot?
- How does the time vary as the height increases?
- Would a simple linear line equation model be a 'good' choice?

1.4.2 Exercise 2: Identifying the Limitations of Linear Models

Objective: Test a simple linear model and identify the limitations of the model.

Implement the following tasks:

- Use the pre-implemented algorithm to model the data with a linear model of the form: $t_i = c_1 + c_2 m_i$.
- Compute the model predictions of the falling times for all weights and heights.
- Compute the accelerations using $a_i = h_i / t_i^2$.
- Visualize the true measurements and the generated predictions using scatter plots.
- Define a suitable error measure and compute it.
- Visualize the original data and the model as colored line within one plot.

Report your results and answer the following questions:

- Can a linear model explain the relationship?
- Why is the linear model insufficient?
- How many linear models could explain the data?
- How would you interpret the model predictions of the accelerations?

1.4.3 Exercise 3: Explore Non-Linear Mixture of Experts Models

Objective: Test a rule based mixture of experts approach using multiple linear models.

Implement the following tasks:

- Develop a decision tree using the height as decision variable (hint: revisit the question on the number of linear models from the previous exercise).
- Implement your decision tree in the pre-implemented decision tree.
- Split the dataset into subset for the individual expert models.
- Train the individual experts models using $d = 1 \dots D$ linear models of the form: $t_i = c_1^d + c_2^d m_i$.
- Compute the prediction errors.
- Visualize the original data and the D models as colored lines within one plot.

Report your results and answer the following questions:

- Does the mixture of experts model improve the prediction quality of the falling times?
- Are the predictions of the accelerations plausible?
- What is the main drawback of this mixture of experts approach?

1.4.4 Exercise 4: Explore Locally Weighted Regression (LWR)

Objective: Understand how locally weighted regression models non-linearities in the data.

Implement the following tasks:

- Explain in your own words how locally weighted regression, illustrated in the provided handout models non-linearities.
- Select for each expert from the previous expert one data sample as test sample.
- Visualize the training and the test data in a scatter plot using difference colors.
- Compute the distances to all training data points for all test points.
- Compute for each test sample the importance weight matrix.

- Predict for each test sample the falling time using the weighted linear regression model.

  Report your results and answer the following questions:

- Comparing the prediction results, how does LWR compare to the mixtures of experts approach?
- How would outlier affect the performance of the mixture of experts and the LWR model?
- How would both approaches perform if a test sample is far off all training samples?

1.5 Jupyter NB Exercise—Introduction to Jupyter

In this exercise, you will take your first steps with Jupyter Notebooks. Our goal is to become familiar with the workflow of Jupyter Notebooks and to understand essential Python basics, such as data types and core scientific libraries.

As a practical example, we will work with thermal expansion measurements for steel, aluminium, and copper at different temperatures. You will load the data, explore it, and visualize it using Python's plotting tools (Fig. 1.4).

1.6 Jupyter NB Exercise—Data Modeling of Objects in Free Fall

In this exercise, we investigate how to model the motion of objects in free fall using real or simulated data. Our goal is to understand how well a linear model can predict physical quantities such as the time an object takes to fall from a given height.

You will begin by constructing a linear regression model and evaluating its predictive performance. Through experimentation, you will discover the limitations

Fig. 1.4 For the latest online version of the notebook, scan the QR code or visit: https://github.com/mul-cps/DMBook/tree/main/ex_introduction

Fig. 1.5 For the latest online version of the notebook, scan the QR code or visit: https://github.com/mul-cps/DMBook/tree/main/ex_linear_models

of linear techniques when describing physical processes governed by non-linear relationships. To address these limitations, you will then explore non-linear modeling approaches that better represent the physics of free fall (Fig. 1.5).

Part II
Processes and Sensors

This part introduces core ideas related to data granularity and the interaction between models and the physical world. We explore different levels of model integration, from static digital models to interactive cyber-physical systems, and examine how sensor characteristics influence data quality. Topics include sensor transfer functions, error propagation, and the computation of prediction bounds. The part concludes with a survey of selected state-of-the-art sensor technologies.

Chapter 2
Processes and Data Granularity

In this chapter, we will explore the critical concept of data granularity, focusing on the appropriate level of detail necessary in data modeling approaches. Determining the right granularity is key to optimizing system performance, especially in processes that rely on sensors and digital models [25].

2.1 Terminology

- **Granule:** A coherent and meaningful unit that encapsulates information. Granules represent chunks or clusters of data that exhibit internal homogeneity and external heterogeneity. Granules help organize and structure complex information.
- **Granularity:** The level or scale at which information is organized, represented, or processed. It pertains to the degree of detail or coarseness in the division of data into granules. Granularity determines the size or extent of the individual granules and how they relate to each other hierarchically.

2.2 Level of Data Granularity

Data granularity refers to the level of detail in data collection and modeling. Depending on the process, different granularities may be more suitable. The right granularity is essential for striking a balance between computational efficiency and the quality of insights drawn from the data. The wrong granularity, either too fine or too coarse, can lead to inefficiencies in data storage, processing, and model accuracy.

E. Rueckert, *Data Modeling*, https://doi.org/10.1007/978-981-95-7592-3_2

2.2.1 Determining Granularity: Key Factors

- **Domain Knowledge:** Understanding the system dynamics, the rate at which changes occur, and the goals of the model.
- **Data Utility:** How much information is retained or lost when moving from a fine-grained to a coarse-grained representation.
- **Storage and Processing Cost:** Higher granularity requires more storage and computational resources.

Mathematically, data granularity in a process model can often be related to the information entropy, which reflects the amount of information captured in the process. A more detailed (fine-grained) model will generally have higher entropy. The information entropy is discussed in detail in Chap. 5.

2.2.2 Time-Dependent Data Granularity

Time-dependent data granularity concerns the temporal resolution at which data is captured. This is critical in time-series data where the sampling rate plays a fundamental role in the system's accuracy and efficiency.

Let's consider a forecasting model for household energy consumption:

- **Too fine-grained (e.g., millisecond sampling):** High storage and computational costs, without much improvement in predictive accuracy.
- **Too coarse-grained (e.g., monthly):** Misses transient behaviors in energy use that could influence predictions.

2.2.3 Process-Based Data Granularity

Process-based granularity refers to the level of detail with which various components and interactions within a system are represented. The decision depends on the process's complexity and the goals of the analysis.

For example, in a multi-stage manufacturing process, the granularity could vary based on the focus of the model:

- **Fine-grained:** Model each machine's internal components (e.g., motors, gears).
- **Coarse-grained:** Model the machines as black boxes with input/output parameters.

2.3 Aggregation and Segmentation Strategies

To achieve the right level of granularity, data aggregation or segmentation techniques are used:

- **Aggregation:** Combining smaller units into a larger one (e.g., aggregating energy consumption from multiple appliances).
- **Segmentation:** Breaking down a process into smaller units to focus on individual parts (e.g., segmenting a production line into its machines).

Granularity can be defined mathematically through hierarchical clustering, where data is recursively grouped based on distance or similarity metrics (e.g., Euclidean distance for physical measurements, cosine similarity for text data).

2.4 Process Models

Process models represent the various interactions and components within a system at different levels of abstraction, from simple digital models to highly integrated cyber-physical systems [2, 20].

2.4.1 Digital Model

A digital model is a static representation of a system or process based on known parameters and equations. It does not incorporate real-time data and is often used for simulations or to study potential outcomes.

For example, a digital model of a wind turbine might simulate its energy production based on wind speed, without being connected to an actual turbine in operation.

2.4.2 Digital Shadow

A digital shadow is an extension of the digital model, wherein real-time data from sensors is used to update the model periodically. This enables the digital representation to reflect the current state of the physical system more accurately, allowing for more timely insights and decisions.

For instance, in the case of a production line, a digital shadow might use sensor data from machines to track wear and tear and predict when maintenance will be required.

2.4.3 *Digital Twin*

The digital twin represents the next level of integration, where the digital model and the physical system operate in sync in real-time. A digital twin not only reflects the current state of a system but also uses predictive models and machine learning algorithms to forecast future states and optimize performance. This is critical for applications such as predictive maintenance, where early detection of potential failures can prevent costly downtime.

In modern industrial applications, digital twins are frequently employed to optimize entire production systems, allowing for real-time decision-making and process improvements.

2.4.4 *Cyber-Physical System*

A cyber-physical system (CPS) is the highest level of integration between digital models and physical systems, where computational algorithms and physical processes are deeply interconnected. In a CPS, sensors collect data from the physical world, which is processed by algorithms, leading to real-time actions or changes in the physical system.

An example of a CPS is an autonomous vehicle, where data from cameras, radars, and other sensors are continuously fed into an AI system that makes real-time decisions to steer, brake, or accelerate.

CPSs are the foundation of modern Industry 4.0 systems, where digitization, data analysis, and automation combine to create highly efficient and adaptable processes.

2.5 Practical Exercises: Processes and Models

To reinforce the concepts discussed in this chapter, this section provides practical exercises focused on modeling processes with varying data granularities, constructing digital shadows, and creating simple digital twins.

- **Exercise 1**: Create a digital model of a household electrical consumption system and determine the appropriate sampling rate for predicting monthly energy costs.
- **Exercise 2**: Build a simple digital shadow using real-time temperature data from a sensor network in a greenhouse. Explore how varying the data granularity affects the accuracy of temperature predictions.
- **Exercise 3**: Design a digital twin for a conveyor belt system in a factory, simulating both real-time performance and predictive maintenance schedules.
- **Exercise 4**: Integrate a cyber-physical system by developing a feedback loop between a sensor-driven model and an automated control system for regulating fluid flow in a pipeline.

Chapter 3
Sensors

While biological sensors like our eyes, ears, our mouth, and our nose help us to understand the world, technical sensors are used in diverse applications ranging from industrial automation to medical diagnostics. In this chapter, we explore the characteristics, noise factors, and technological implementations of sensors.

3.1 Sensor Characteristics

This section starts with definitions of key terminologies used to describe sensors. Thereafter, commonly used sensor categorizations strategies and essential sensor characteristics are discussed.

3.1.1 Terminology

To evaluate and compare sensors, several fundamental characteristics are commonly used. These properties describe how a sensor behaves under ideal and non-ideal conditions and directly influence the quality of the measured data and subsequent model predictions.

- **Accuracy**: The degree to which a sensor's measurement conforms to the true value of the process variable p.
  Example: A temperature sensor reading 25.1 °C for a true temperature of 25.0 °C exhibits high accuracy.
  Note: Accuracy is affected by both systematic errors (bias) and random measurement errors.
- **Precision**: The consistency or repeatability of a sensor's measurements under unchanged conditions, independent of the true value.

E. Rueckert, *Data Modeling*, https://doi.org/10.1007/978-981-95-7592-3_3

Example: A pressure sensor that repeatedly reports 101.5 kPa over multiple measurements demonstrates high precision, even if the true pressure is 100 kPa.
Note: A sensor can be precise but not accurate if its measurements are consistently biased.

- **Sensitivity**: The ability of a sensor to detect small changes of the value being measured.
Example: A load cell capable of resolving weight changes as small as 0.01 g has high sensitivity.
Note: Sensitivity is commonly quantified as the change in sensor output per unit change in input (e.g., mV/°C for a temperature sensor).
- **Range**: The interval of input values over which a sensor can measure the process variable reliably and within specified accuracy limits.
Example: A humidity sensor with a specified range from 0% to 100% relative humidity.

These characteristics must be considered jointly when selecting sensors for a given application, as improvements in one property (e.g., sensitivity) may come at the cost of others (e.g., range or robustness).

3.1.2 Categorizations

Various sensor categorizations can be used to describe and organize the numerous sensing devices, including categorizations related to the **detection means used in sensors** (e.g. biological, chemical, electrical, magnetic or electromagnetic waves, heat, temperature, mechanical displacement, radioactive waves, radiation or other) or related to their **stimulus** (e.g. acoustic, chemical, electrical, magnetic, optical, mechanical, radiation or heat). In this chapter, the term stimulus is represented as a **process variable** p, which reflects the measurable quantity or condition being observed by the sensor.

Here are examples of sensor technologies categorized by the type of **stimulus** they detect:

- **Acoustic Sensors**: Measure sound waves and vibrations (e.g., microphones, ultrasonic sensors).
- **Chemical Sensors**: Detect chemical substances, such as gases or liquids, commonly used in environmental monitoring (e.g., gas sensors, pH sensors).
- **Electrical Sensors**: Detect electrical changes or properties, typically used in circuits and automation (e.g., voltage sensors, current sensors).
- **Magnetic Sensors**: Sense magnetic fields and their variations (e.g., Hall effect sensors, magnetometers).
- **Optical Sensors**: Measure light intensity, color, or radiation, often applied in imaging or distance measurement (e.g., photodiodes, LIDAR, RGB cameras, depth cameras).

- **Mechanical Sensors**: Measure forces, displacement, pressure, or vibration (e.g., gyroscopes, accelerometers, pressure sensors).
- **Radiation Sensors**: Detect radiation levels, frequently used for safety and health monitoring (e.g., Geiger counters, dosimeters).
- **Thermal Sensors**: Measure heat or temperature changes in a system (e.g., thermocouples, thermistors).

For further discussions and for more detailed categorical classifications, the *Handbook of Modern Sensors* [9] is recommended.

3.1.3 Sensor Transfer Function

All sensors are inherently affected by noise and uncertainties stemming from various sources such as environmental conditions, sensor inaccuracies, discretization errors, hysteresis, or unmodeled non-linearities. Mathematically, this behavior can be modeled as:

$$s = f(p) + \epsilon \ , \tag{3.1}$$

- p is the physical process variable,
- s is the observed sensor output (digital signal),
- $f(p)$ is the non-linear transfer function describing the sensor's response,
- ϵ represents noise, which may include sensor-specific errors and external factors.

Remark 3.1 The transfer function f encapsulates not only the sensor's measurement but also the analog-to-digital (A/D) conversion, transmission, and pre-processing steps. Therefore, we treat sensor systems as encapsulated entities that convert physical quantities into digital signals.

3.1.4 Reconstruction and Inverse Transfer Function

In practice, we often need to reconstruct the original physical variable p from the noisy sensor data. This is achieved through the inverse transfer function:

$$\tilde{p} = f^{-1}(s - \epsilon) \ , \tag{3.2}$$

where $\tilde{p}$ is the reconstructed estimate of p, and $f^{-1}(.)$ is the inverse of the sensor transfer function in Eq. 3.1. In an ideal scenario, with no noise ($\epsilon = 0$), the physical variable can be perfectly recovered:

$$p = f^{-1}(s) \ . \tag{3.3}$$

However, due to inevitable noise and uncertainties, this reconstruction is imperfect, leading to errors that depend on factors like the accurarcy of the sensor technology, operating conditions, and system non-linearities.

3.1.5 Reconstruction Error

To quantify the reconstruction accuracy, the error e between the original physical variable p and the reconstructed variable $\tilde{p}$ can be measured using for example the squared Euclidean distance (L2-norm):

$$e = (p - \tilde{p})^2 \ = ||p - \tilde{p}||_2^2 \, .$$

The symbol $|| \, . \, ||_2^2$ is used to denote the squared L2-norm, where $|| \, . \, ||_2$ is used for the classical (non-squared) L2-norm.

For cases involving multiple variables $\boldsymbol{p} = [p_1, p_2, \ldots, p_D]$, the reconstruction error is:

$$e = (\boldsymbol{p} - \tilde{\boldsymbol{p}})^T \ (\boldsymbol{p} - \tilde{\boldsymbol{p}}) \ ,$$

where e is a scalar, representing the total squared error across all dimensions.

3.1.6 Linearizations and Assumptions

In many cases, sensor models are linearized around specific operating points (e.g., room temperature). This linearization simplifies the inverse function $f^{-1}(.)$, making it more manageable for analysis and computation. However, it introduces assumptions, and errors grow as the sensor operates further from the linearization or operating point [9].

Remark 3.2 The back transformation of any sensor variable requires to model the inverse transfer function $f^{-1}(.)$ to compute the original physical process variable. However, all reconstructions based on $f^{-1}(.)$ are subject to errors due to noise, nonlinearities, and approximations.

3.2 Measurement Errors and Model Prediction Bounds

When building data models using sensor data, it's critical to understand the potential errors and upper bounds of predictive performance. These insights help prevent overfitting, guide optimization, and ensure that model predictions are not overestimated. Below, we discuss common strategies for quantifying uncertainty in sensor-based predictions.

3.2.1 *Uncertainty of the Predictive Variable $\tilde{p}$*

In many cases, uncertainties are provided directly for the target variable $\tilde{p}$, representing the physical quantity being measured.

For instance, consider a temperature sensor that estimates temperature based on resistance changes, such as a thermistor. If the sensor's datasheet specifies an accuracy of ±0.2 °C, then any predictive model using this sensor will have a minimum expected error of 0.2 °C.

Remark 3.3 Scenarios where uncertainty is directly tied to the target variable are common. Examples include the prediction of future measurements given historic data, e.g., predicting temperature curves or energy consumption curves. Other examples aim at replacing 'costly' high-precision sensors with low-cost alternatives, e.g., temperature prediction problems based on datasets of numerous low-cost sensor readings as inputs and 'costly' high-precision measurements as targets. After the model was trained from the data, the 'costly' high-precision device is replaced by the trained model.

3.2.2 *Uncertainty of the Measureing Process s*

Another common scenario aims at computing performance bounds given uncertainties in the input variables of data modeling approaches. To convert this input uncertainty into the space of the predictive variable $\tilde{p}$ several strategies were proposed. All these strategies assume some *linearity* or *convexity* of the inverse transfer function in Eq. 3.2.

3.2.2.1 Classical Error Propagation

Assuming a linear relationship between $\tilde{p}$ and multiple inputs s_i with $i = 1 \ldots D$, the total uncertainty $\Delta\tilde{p}$ can be approximated using a first-order Taylor expansion:

$$\Delta\tilde{p} = \frac{\delta\tilde{p}}{\delta s_1}\Delta s_1 + \frac{\delta\tilde{p}}{\delta s_2}\Delta s_2 + \ldots + \frac{\Delta\tilde{p}}{\delta s_D}\delta s_D.$$

This first-order approximation assumes linear dependencies between the target variable and the inputs.

3.2.2.2 Gaussian Error Propagation Law

In cases where input uncertainties follow a Gaussian distribution, Gaussian error propagation can be computed.

For each sensor input s_i uncertainties assuming Gaussian noise, i.e., $s_i \sim \mathcal{N}(s_i|0, \sigma_i)$ are given. The symbol $\mathcal{N}(.)$ denotes the Gaussian (univariate or one-dimensional) distrubution:

$$\mathcal{N}(\mu|0, \sigma) = \frac{1}{(2\pi\sigma^2)^{1/2}} \exp^{\{-\frac{1}{2\sigma^2}(x-\mu)^2\}} . \tag{3.4}$$

The Gaussian error propagation differs between four basic algorithmic operations:

- **Addition and Subtraction:** For $\tilde{p} = s_1 + s_2 - s_3$, the resulting target uncertainty is give by

$$\sigma_{\tilde{p}}^2 = \sigma_1^2 + \sigma_2^2 + \sigma_3^2.$$

- **Multiplication and Division:** For $\tilde{p} = s_1 s_2 / s_3$, the resulting target uncertainty is given by

$$\left(\frac{\sigma_{\tilde{p}}}{\tilde{p}}\right)^2 = \left(\frac{\sigma_1}{s_1}\right)^2 + \left(\frac{\sigma_2}{s_2}\right)^2 + \left(\frac{\sigma_3}{s_3}\right)^2 .$$

This method is only valid for linear systems, so non-linear cases require more advanced techniques, such as Jensen's inequality, discussed next.

3.2.2.3 Jensen's Inequality Approach

When dealing with non-linear systems, Jensen's inequality can be applied to derive an upper bound on the error. Based on the inverse transfer function $f^{-1}(.)$, the error bounds are computed as:

$$\begin{aligned} p &= f^{-1}(s) . \\ \tilde{p} &= f^{-1}(s+\epsilon) , \\ &\leq f^{-1}(s) + f^{-1}(\epsilon) . \\ \tilde{p} - p &\leq f^{-1}(\epsilon) . \end{aligned}$$

The term $f^{-1}(\epsilon)$ provides an upper bound for the maximum of the error of the reconstruction. This bound is based on *Jensen's Inequality*, where

$$\mathrm{e}\{ f^{-1}(x+y) \} \leq f^{-1}(\mathrm{e}\{x\}) + f^{-1}(\mathrm{e}\{y\}) .$$

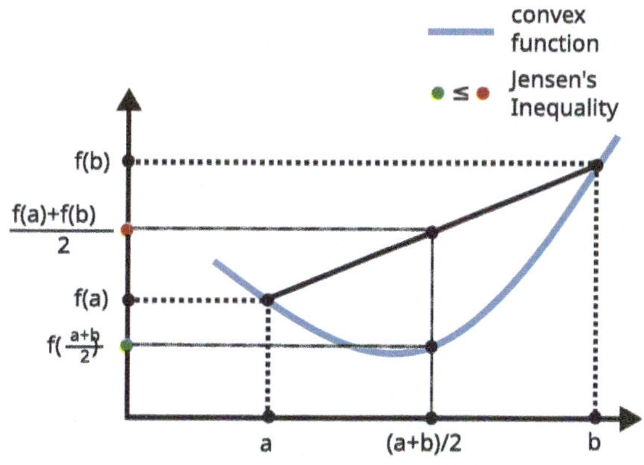

Fig. 3.1 Jensen's Inequality states that $f\left(\frac{a+b}{2}\right) \leq \frac{f(a)+f(b)}{2}$, which holds if f is a convex function, as illustrated in this sketch

The symbol e{.} denotes the expection, or in other words the avarge expected value. Typically the expectation is computed via the (weighted) *mean* [10].

The above upper bound is based on *Jensen's Inequality* [11], which assumes a **convex function** $f^{-1}(.)$. A visual representation of Jenson's inequality is shown in Fig. 3.1.

Example 3.1 The uncertainty of the target value $\tilde{p}$ can be calculated from the sensor noise ϵ via $f^{-1}(\epsilon)$. A temperature sensor could, for example, convert a voltage signal into temperatures using a quadratic (i.e. convex) function, $f^{-1} = x^2$. For a given noise $\epsilon = 0.1$, the target uncertainty would be 0.01.

Remark 3.4 Note that Jensen's inequality also holds for numerous variants like nested convex function, e.g., $f^{-1} = f^{-1}(g(s))$, where both functions, $f^{-1}(.)$ and $g(.)$ are convex.

3.3 Sensor Technologies

In this section, we briefly present several common sensor technologies. Each subsection discusses the *stimuli*, *physical principles*, assumptions regarding *sensing accuracy*, and *operating conditions*.

In addition, product examples are presented with their specifications of the working conditions and selected features relevant for data modeling.

3.3.1 Acoustic Sensors

Stimuli Sound waves (pressure variations in the air).

Physical Principles Acoustic sensors convert sound waves into electrical signals through mechanisms such as piezoelectricity or changes in capacitance. A common

Fig. 3.2 Illustration of the microphone STMicroelectronics MP23DB01HP (Source: STMicroelectronics) [30]

type is the microphone, where sound pressure induces mechanical vibrations that are converted to electrical signals.

Assumptions Acoustic sensors are highly sensitive to environmental noise and reverberations, which can affect their accuracy. They generally work best in controlled environments or with noise-canceling techniques.

Example STMicroelectronics MP23DB01HP Microphone (Fig. 3.2).

- **Operating Range:** The temperature operation range is from −40 °C to +85 °C. The microphone can function effectively from a low SPL (quiet sound) up to an acoustic overload point (AOP) of 135 dB.
- **Sensitivity:** It has a sensitivity rating of −42 to −40 dBFS, which refers to the microphone's response to a standard input sound. The sensitivity is measured at 1 kHz, 94 dBSPL.
- **Noise:** It offers a signal-to-noise ratio (SNR) of 65 dB(A), which indicates how well the microphone can capture the desired sound while minimizing background noise. The total harmonic distortion (THD) remains low, around 0.2% at 94 dB, and increases to 0.5% at higher SPLs of 110 dB.
- **Usage:** It's ideal for applications where accurate audio data is needed, such as voice recognition and environmental sound monitoring in both industrial and consumer electronics.
- **Link to Datasheet:** You can explore more details about the microphone here [30].

3.3.2 *Chemical Sensors*

Stimuli Chemical concentrations or reactions.

Physical Principles Chemical sensors detect the presence or concentration of specific chemicals of chemical reactions that generate an electrical signal. Examples include pH sensors, gas sensors, and biosensors (Fig. 3.3).

Assumptions These sensors rely on consistent chemical interactions and are often influenced by temperature, humidity, and pressure. Calibration is critical for accurate readings.

Fig. 3.3 Illustration of the chemical sensor MQ-135 (Source: SIMAC ELECTRONICS) [19]

Example MQ-135 Gas Sensor.

- **Operating Range:** The MQ-135 sensor can detect a wide range of gases, including NH3, NOx, alcohol, benzene, CO2, and smoke. Its operating temperature range is from −20 °C to +50 °C, and it operates effectively in humidity levels ranging from 20% to 90%.
- **Sensitivity:** The sensor has a measurement range of 10–1000 ppm, depending on the gas. For ammonia (NH3) and alcohol it is 10–300 ppm, for benzene the range is 10–1000 ppm.
- **Noise:** The MQ-135 can exhibit some noise under fluctuating environmental conditions (temperature and humidity). Calibration is typically necessary to minimize false positives, especially in complex environments. Rs in air/Rs(100 ppm NH3) $\geq$5.
- **Usage:** This sensor is used in air quality monitors, detecting harmful gas concentrations in indoor environments, industrial safety, and even smart agriculture to monitor air pollutants.
- **Link to Datasheet:** You can explore more details about the sensor here [19].

3.3.3 Electromagnetic Wave Sensors

Stimuli Electromagnetic waves, typically in the microwave, radio, infrared (IR), or millimeter-wave spectrum.

Physical Principles Electromagnetic wave sensors detect and measure electromagnetic radiation by emitting waves (for example, microwaves) and analyzing the reflected signal. Common types include radar sensors, infrared sensors, and microwave sensors. These sensors utilize phenomena such as Doppler shifts to detect motion or the time-of-flight principle to measure distance. Reflections are processed to infer properties such as speed, position, or presence of an object.

Assumptions These sensors assume the absence of significant interference from other electromagnetic signals and often require a clear line of sight. Accuracy can be affected by environmental factors such as rain, dust, or multiple reflections in cluttered environments.

Fig. 3.4 Illustration of the XENSIV™ radar sensor BGT24MTR11 (Source: Infineon Technologies) [17]

Example Infineon BGT24MTR11 24 GHz Radar Sensor (Fig. 3.4).

- **Operating Range:** This sensor operates over a frequency range of 24 GHz with an operating temperature from −40 °C to +125 °C. Its maximum range is up to 20 m.
- **Sensitivity:** The sensor is highly sensitive to small motions, able to detect objects moving as slowly as a few centimeters per second.
- **Noise:** The signal-to-noise ratio (SNR) of the sensor depends on distance and environmental conditions. It uses built-in filtering to minimize noise, especially in indoor environments.
- **Usage:** Ideal for occupancy detection, industrial automation, and vehicle radar systems.
- **Link to Datasheet:** You can explore more details about the radar sensor here.

3.3.4 Flow Sensors

Stimuli Fluid or gas flow.

Physical Principles Flow sensors typically use either mechanical methods (e.g., turbine flow meters) or thermal methods (e.g., hot-wire anemometers) to measure the velocity and volume of fluid or gas passing through them.

Assumptions Turbulence, viscosity changes, and temperature variations can impact accuracy. Many flow sensors assume steady-state flow conditions for optimal performance.

Example Honeywell AWM720P1 Microbridge Mass Airflow Sensor.

- **Operating Range:** The Honeywell AWM720P1 can detect mass airflow rates from 0 to 200 standard liters per minute (SLPM). It operates in a temperature range of −25 °C to +85 °C and can function in a relative humidity range from 0% to 100%, non-condensing.
- **Sensitivity:** It has high sensitivity to small flow rates, with a fast response time of 1 millisecond. The sensor outputs an analog voltage that varies linearly with the airflow rate, allowing for precise measurements in various low-flow applications.

- **Noise:** The sensor provides low-noise readings due to its microbridge structure, ensuring consistent airflow measurement. It also has built-in filtering to reduce environmental noise effects, such as vibrations.
- **Usage:** This flow sensor is ideal for applications in medical equipment (e.g., ventilators, anesthesia delivery systems), HVAC systems, and gas leak detection systems. It is also used in industrial automation where precise airflow monitoring is required.
- **Link to Datasheet:** You can explore more details about the sensor here [15].

3.3.5 Force, Strain, Pressure and (Non-optical) Tactile Sensors

Stimuli Mechanical forces (pressure, tension, compression).

Physical Principles These sensors use piezoelectric, capacitive, or resistive methods to measure mechanical forces. Strain gauges measure deformation in materials, while tactile sensors, often used in robotics, detect pressure distributions.

Assumptions These sensors are affected by material fatigue, temperature changes, and the quality of mechanical contact. They assume linear elastic behavior within the sensing material.

Example Honeywell FSS Series Force Sensor.

- **Operating Range:** The Honeywell FSS Series sensor operates in a force range of 0 to 1500 grams (g) with a maximum overforce limit of 4.5 kg. Its operating temperature ranges from $-40\,°C$ to $+85°C$, making it suitable for a variety of industrial and medical environments.
- **Sensitivity:** The sensor offers repeatablity of ± 10 g at 300 g. Further, a sensitivity of 0.14 mV per gram is specified in the datasheet. It responds (almost) linearly to applied force with a linearity (B.FS.L.) of $\pm 1.5\%$ span.
- **Noise:** Due to its robust design, the sensor exhibits low noise and is able to maintain a high signal-to-noise ratio even in high-force environments. It is also resistant to electromagnetic interference (EMI), ensuring stable and reliable readings.
- **Usage:** The Honeywell FSS force sensor is widely used in medical applications (e.g., infusion pumps, dialysis machines), industrial load monitoring, and robotic systems requiring precise force feedback.
- **Link to Datasheet:** You can explore more details about the sensor here [16].

3.3.6 Humidity, Temperature and Moisture Sensors

Stimuli Moisture or humidity levels, or thermal energy (temperature).

Fig. 3.5 Illustration of the Sensirion SHT41 humidity and temperature sensor (Source: SENSIRION AG) [28]

Physical Principles Humidity sensors often use capacitive or resistive methods to measure the amount of water vapor in the air. Changes in humidity alter the dielectric constant of materials, producing measurable changes in capacitance or resistance.

Temperature sensors measure heat levels through thermocouples, resistive temperature devices (RTDs), or infrared detection. Each sensor type relies on changes in material properties (e.g., electrical resistance, voltage) to infer temperature.

Assumptions Accuracy depends on stable temperature and pressure conditions. Many sensors assume a constant relationship between humidity and dielectric properties.

Example Sensirion SHT41 Humidity and Temperature Sensor (Fig. 3.5).

- **Operating Range:** The Sensirion SHT41 operates in a humidity range from 0% to 100% relative humidity (RH) and a temperature measurement range of 0 °C to 90 °C. The sensor operates at a temperature from −40 °C to +125 °.
- **Sensitivity:** This sensor provides high accuracy, with a typical humidity accuracy of ±2% RH and a temperature accuracy of ±0.2 °C. The response time for humidity is 8 seconds, and for temperature it is more than 2 seconds. Thus, fast changes cannot be captured.
- **Noise:** The sensor boasts a low noise floor with a signal-to-noise ratio that minimizes interference from environmental factors. Its digital output (I^2C or analog) helps reduce external noise, ensuring accurate and stable measurements.
- **Usage:** The SHT41 is used in various applications, such as smart home systems, consumer electronics, where both temperature and humidity monitoring are critical but rate of changes are slow. Its small form factor and low power consumption make it ideal for embedded systems and IoT applications.
- **Link to Datasheet:** You can explore more details about the sensor here [28].

3.3.7 Light Detectors

Stimuli Visible, infrared, or ultraviolet light.

Physical Principles Light detectors convert photons into electrical signals, often using the photoelectric or photoconductive effects. Various types of light detectors include:

- **Photodiodes:** These devices detect light by generating a current when exposed to photons. They are sensitive to a broad spectrum, including visible and infrared light, and are often used in optical communication systems, smoke detectors, and ambient light sensors.
- **CMOS Sensors (Complementary Metal-Oxide-Semiconductor):** These are widely used in RGB cameras, which convert light into an electronic signal for image capture. CMOS sensors are faster and consume less power than CCD sensors, making them ideal for mobile devices and embedded systems. However, they can suffer from rolling shutter effects, where fast-moving objects appear distorted.
- **IR Cameras (Infrared Cameras):** These cameras detect infrared radiation, often used in low-light or thermal imaging applications. They can be used for night vision, heat detection, or object tracking. IR cameras are often integrated into security systems and industrial monitoring.
- **Depth Cameras (RGB-D Sensors):** Depth cameras, such as the Intel RealSense D435i, use structured light or time-of-flight (ToF) principles to measure the distance between the camera and objects in the environment. The D435i combines an RGB camera with an infrared depth sensor to create 3D models of objects and scenes. It operates effectively in various lighting conditions, providing depth data along with visual RGB information. The Intel RealSense D435i has a depth accuracy range of up to 10 meters, with a global shutter IR camera that minimizes motion blur and distortion.
- **Optical Tactile Sensors:** These sensors use light-based methods to detect touch and pressure. For example, the DIGIT sensor uses LEDs and a camera to capture deformations in a translucent surface, allowing it to measure contact forces and tactile patterns. Optical tactile sensors offer high resolution for fine tactile sensing, making them suitable for robotic manipulation tasks.

Assumptions Light detectors generally assume direct line-of-sight to the light source and a stable light environment. External light interference, ambient light variations, and occlusions can degrade accuracy, particularly for systems relying on structured light or time-of-flight methods (Fig. 3.6).

Example Intel RealSense D435i Depth Camera.

- **Operating Range:** The Intel RealSense D435i can measure depth within a range of 0.28 to 3 meters. Its depth camera operates in indoor and outdoor conditions, though bright sunlight can affect depth accuracy.
- **Sensitivity:** The depth accuracy decreases as the distance increases. At 2 meters, the typical accuracy is $<2\%$. The RGB camera captures color images at 1080p

Fig. 3.6 Illustration of the Intel RealSense D435i sensor (Source: E. Rueckert, TU Leoben, Austria) [18]

resolution with a Field of View (FOV) of 69° × 42°. The Depth FOV is limted to 87° × 58°. The frame rates are between 30 and 90 frames per second (fps).
- **Noise:** The camera has a depth noise variance of about 2% at shorter distances, increasing to 5% at maximum range. Motion artifacts are minimized with the global shutter on the IR camera.
- **Usage:** The D435i is ideal for robotics, virtual reality, and 3D scanning, providing accurate depth data for object detection, obstacle avoidance, and scene reconstruction. It includes an onboard inertial measurement unit (IMU) for tracking motion in 3D.
- **Link to Datasheet:** You can explore more details about the camera here [18].

3.3.8 Velocity Sensors

3.3.9 Acceleration Sensors

Stimuli Changes in velocity and acceleration.

Physical Principles Velocity and acceleration sensors (accelerometers) measure changes in an object's speed or movement by detecting inertial forces. These sensors typically work using piezoelectric, capacitive, or MEMS (Micro-Electro-Mechanical Systems) technologies, where mechanical deformation or capacitance changes generate electrical signals corresponding to the applied force or acceleration. Velocity can be inferred from integrated acceleration data.

Assumptions These sensors assume minimal external vibrations and interference. External factors like temperature and electromagnetic noise can affect sensor precision. Proper calibration is often necessary to ensure accurate velocity and acceleration readings.

Example Bosch Sensortec BMA400 Accelerometer (Fig. 3.7).

Fig. 3.7 Illustration of the Bosch Sensortec BMA400 Accelerometer (Source: Bosch Sensortec) [4]

- **Operating Range:** The BMA400 operates with a measurement range of ± 2 g, ± 4 g, ± 8 g, or ± 16 g and a temperature range of $-40\,^{\circ}\mathrm{C}$ to $+85\,^{\circ}\mathrm{C}$. Its ultra-low power consumption makes it suitable for always-on applications.
- **Sensitivity:** The sensor has a sensitivity of up to 4096 LSB/g (Least Significant Bits per g-force), allowing it to detect fine accelerations and movements.
- **Noise:** It boasts a low noise density of $150\,\mu\mathrm{g}$ per $\mathrm{Hz}^{1/2}$, making it ideal for detecting small accelerations with minimal external noise interference.
- **Usage:** Commonly used in smart devices, wearables, and motion detection applications, it is particularly well-suited for gesture recognition, fall detection, and fitness tracking.
- **Link to Datasheet:** You can explore more details about the accelerometer here [4].

3.4 Practical Exercises: Sensor Uncertainties and Upper Bounds of Predictive Models

3.4.1 Exercise 1: Selecting Sensor Technologies

Objective: Your goal is to classify Gummi Bears according to the features color, weight and height. Learn how to select adequate sensor technologies.

Perform the following tasks:

- Research and list at least three different sensor technologies suitable for measuring color, weight, and height of Gummi Bears.
- For each selected sensor, retrieve the datasheet and summarize the key specifications, including resolution, sensitivity, operational range, and noise levels.
- Create a comparison table highlighting the strengths and weaknesses of each sensor technology concerning the specific application of classifying Gummi Bears.
- Discuss how each sensor's characteristics could affect the accuracy of the classification task.

Report the results. Answer the following questions:

- Which sensor technology would you recommend for each feature (color, weight, height) and why?
- How do the specifications of the selected sensors impact the overall classification performance?
- What factors should be considered when integrating multiple sensors into a single system?

3.4.2 Exercise 2: Best Achievable Prediction Results in a Single-Sensor System

Objective: Given is a single sensor technology with known properties. Given is the uncertainty of the measureing process s. Test different error propagation strategies to compute upper performance bounds. Further, understand how to extract uncertainty-relevant information from datasheets and compute upper performance bounds for simple linear regression methods.

Perform the following tasks:

- Choose one of the sensors identified in Exercise 1 and analyze its datasheet to identify key uncertainty parameters (e.g., accuracy, precision, and noise levels).
- Compute the upper performance bounds for a linear regression model that predicts one of the features (e.g., weight) based on the sensor readings. Use the identified uncertainties to inform your calculations.
- Apply Jensen's inequalty to transform the sensor uncertainty into the target variable space.
- Compare the results of the classical error propagation law to the strategy using Jensen's inequalty.

Report the results. Answer the following questions:

- What uncertainties did you identify from the datasheet, and how do they affect the prediction?
- Which strategy, the classical error propagation law or Jensen's inequality approach, performed best?
- What happens if f^{-1} is a non-linear function?

3.4.3 Exercise 3: Gaussian Error Propagation Law

Objective: Compute upper performance bounds of prediction tasks using the Gaussian error propagation law for a multiple sensor systems.

Fig. 3.8 For the latest online version of the notebook, scan the QR code or visit: https://github.com/mul-cps/DMBook/tree/main/ex_sensors

Perform the following tasks:

- Select at least two sensors from your previous exercises that could work together to improve classification accuracy (e.g., an RGB camera for color and a load cell for weight).
- Research how data from multiple sensors can be combined, focusing on the Gaussian error propagation law.
- Identify the uncertainties associated with each selected sensor and analyze how these might interact when combined.
- Compute the overall upper performance bounds for a multi-sensor system predicting the classification of Gummi Bears based on integrated sensor data.
- Discuss any potential challenges and advantages of using a multi-sensor approach, including considerations of complexity and data processing requirements.

Report the results. Answer the following questions:

- How do the uncertainties from each sensor contribute to the overall predictive performance of the multi-sensor system?
- What strategies can be employed to mitigate the impact of sensor noise when integrating data from multiple sensors?
- Based on your analysis, what conclusions can you draw about the feasibility of achieving improved classification performance with a multi-sensor approach?

3.5 Jupyter NB Exercise—Sensors

In this exercise, we explore two fundamental aspects of sensing in data-driven systems. First, you will analyze measurement uncertainty and error propagation for a single sensor and learn why understanding noise characteristics is essential for reliable data acquisition. Second, you will examine how multiple sensors can be combined to estimate target variables more accurately, illustrating the principles behind multi-sensor systems and sensor fusion.

Through hands-on experiments and simulations, you will gain insight into how sensor characteristics influence downstream model performance and why real-world perception systems rarely rely on one measurement source alone (Fig. 3.8).

Part III
Data and Information Theory

In this part, we introduce the concepts of objects and attributes, distinguish data objects from data types, and survey data storage systems—from relational databases to cloud platforms—while also covering key ideas from information theory, including entropy as a measure of uncertainty and the Shannon—Nyquist sampling theorem. We further review core analysis practices: basic statistical concepts, effective data visualization, normalization techniques, outlier detection, and approaches to measuring data similarity.

Chapter 4
Data

This chapter delves into various data-related concepts, including data types, discretization, and data storage methods. It also explores the information-theoretic aspects of data, such as entropy and redundancy, which play a critical role in understanding data efficiency, compression, and loss.

4.1 Objects and Attributes

In data modeling, objects (also known as instances or entities) represent individual units of data, while attributes (or features) represent the properties or characteristics of these objects. The relationship between objects and their attributes is foundational to structuring, storing, and processing data.

4.1.1 Relationships and Visualization Techniques

Data relationships describe how objects are connected or interact with one another. These relationships can be one-to-one, one-to-many, or many-to-many and are often visualized through diagrams. Three primary techniques are discussed below:

4.1.1.1 Entity-Relationship Diagrams (ERD)

ER Diagrams model relationships between different entities in an object-attribute representation, such as class models or databases. They visually represent entities (objects), attributes (columns), and relationships (associations or keys). In database design creation, ERD are essential to eliminate redundancies.

E. Rueckert, *Data Modeling*, https://doi.org/10.1007/978-981-95-7592-3_4

4.1.1.2 Universal Modeling Language (UML) Diagrams

UML Diagrams are used for object-oriented design and modeling. They represent objects, classes, and the relationships between them. UML diagrams help visualize the structure of complex systems, illustrating attributes, operations, and inheritance, which are useful for software design as well as complex database systems.

4.1.1.3 Graph Database Visualizations

Graph databases store information as nodes (objects) and edges (relationships). Visualizations in graph databases, like *Neo4j* or *Dgraph* [32], depict how entities interact through interconnected relationships. These visualizations are highly useful in understanding complex, non-hierarchical relationships, such as social networks or transportation networks.

4.2 Data Objects and Data Types

Data objects and types play a crucial role in determining how data is stored, manipulated, and processed.

4.2.1 Data Objects

Data objects represent structured entities with multiple attributes. For example, in a database of students, each student is a data object with attributes such as name, age, and enrollment number. Data objects abstract real-world entities and are essential in both database management and object-oriented programming.

The reuse of structures (through classes and objects in programming languages) helps in avoiding redundancy and makes the data more manageable.

4.2.2 Data Types

Data types define the nature of values that can be stored in a variable. They help systems understand how to handle and store the data efficiently.

- **Primitive Data Types:**
  - **Boolean:** Represents two values (true or false).
  - **Character:** A single character (e.g., ‘a’, ‘Z’).
  - **Integer:** Whole numbers (e.g., 1, −100, 2000).

  - **Floating-Point:** Numbers with decimal points (e.g., 3.14, −0.001).
  - **Undefined:** undefined variables (e.g., NaN, Null).

- **Non-Primitive Data Types:**
  - **Strings:** A sequence of characters, e.g., "Hello, World".
  - **Arrays and Tensors:** Multi-dimensional structures for numerical data storage, especially common in scientific computing and machine learning.
  - **Lists, Sets, and Dictionaries:** Commonly used in programming to store collections of elements. Lists are ordered, sets are unordered with unique elements, and dictionaries store key-value pairs.

4.3 Data Storage

Data storage is a critical aspect of data management, dictating how and where data is kept, retrieved, and managed over time.

4.3.1 Databases

Databases are essential for structuring and storing data efficiently, allowing for easy retrieval and management.

4.3.1.1 Relational Databases

Relational databases use tables to store data and enforce relationships through foreign keys and indexes. The most commonly used relational databases include:

- **SQL Databases:** Examples include Microsoft SQL Server, MariaDB, MySQL, and SQLite. These databases use Structured Query Language (SQL) for data management and querying.
- **Time-Series Databases:** Designed for time-stamped data, these databases (e.g., InfluxDB, TimescaleDB) are optimized for scenarios where the data grows over time, such as in sensor data logging or stock price monitoring.

4.3.1.2 Object Databases

Object databases (e.g., ObjectDB) [7, 31] store data as objects rather than relational tables. These databases are ideal for applications built around object-oriented programming languages like Java or C++.

4.3.1.3 Graph Databases

Graph databases [32], store data in a graph format, using nodes (entities) and edges (relationships). These are highly efficient for traversing relationships in complex networks, like social graphs or recommendation engines.

4.3.2 Document-Based Data Storage

Document-based storage systems are non-relational databases that store data as documents, often in JSON-like formats. They are particularly useful in web applications where data can vary in structure.

4.3.2.1 Text Files

Text files like CSVs or JSON files are commonly used for storing tabular or semi-structured data. Though lightweight and easy to manage, they are not suitable for large datasets due to limited querying capabilities and lack of indexing.

4.3.2.2 Spreadsheets

Spreadsheets (e.g., Microsoft Excel, Google Sheets) provide a simple, visual way to manage small to medium datasets. They come with built-in tools for filtering, sorting, and plotting. However, spreadsheets struggle to handle large datasets or complex relationships due to memory constraints and limited functionality for data integrity enforcement.

4.4 Cloud Data Management

Modern cloud-based data management systems have revolutionized the way organizations store, access, and collaborate on data. By enabling data access from anywhere at any time, cloud solutions allow multiple users to work collaboratively on shared documents, breaking down geographical barriers.

Beyond simple storage, cloud platforms offer advanced capabilities such as online document editing, automated backups, disaster recovery, and seamless integration with analytics tools, all while maintaining high levels of security and compliance. In this section, two major strategies to use such data management systems are discussed.

4.4.1 World's Large-Scale Data Storage Centers

Large-scale data centers, often managed by tech giants, store and process petabytes of data. Examples include:

- **Google Data Centers:** Handling search queries, email, and cloud services.
- **Amazon Web Services (AWS):** Providing cloud computing and storage services to businesses worldwide.
- **Microsoft Azure:** Cloud-based services for storing and processing large-scale data.

These data centers often employ distributed storage architectures, redundancy strategies, and advanced cooling mechanisms to maintain massive datasets while ensuring availability and fault tolerance.

However, users have no control over where their data is physically stored, as the storage location is managed by the service providers. This lack of transparency requires users to trust these companies to handle their data securely, adhere to compliance regulations, and safeguard sensitive information from breaches.

4.4.2 Self-Hosted Data Storage Centers

As cloud storage solutions become increasingly dominant, an attractive alternative for organizations that prioritize data control, privacy, or cost efficiency is to implement self-hosted data storage systems. This option provides the flexibility to store, manage, and control data locally or on private cloud infrastructure.

4.4.2.1 Advantages of Self-Hosted Systems

- **Data Privacy and Control:** One of the primary reasons organizations opt for self-hosting is complete control over sensitive data. Unlike public cloud providers, where data might be subject to external policies, in a self-hosted setup, the organization fully governs access, security, and compliance.
- **Customization:** Self-hosted solutions can be tailored to meet the specific needs of the organization. This includes configuring storage architecture, backup processes, and access control systems in ways that public cloud solutions may not offer.
- **Cost Efficiency:** For organizations with significant data storage needs, self-hosted systems can be more cost-effective in the long run. While there is an upfront cost in hardware and setup, ongoing operational costs may be lower compared to paying recurring fees for cloud services, especially when managing large datasets.

4.4.2.2 Tools and Technologies for Self-Hosting

- **Nextcloud:** One of the most popular self-hosted file-sharing and collaboration platforms, Nextcloud provides a private cloud solution. It allows for features like online document editing, file synchronization, and secure sharing. Nextcloud can be deployed on personal servers, ensuring that data remains within the organization's control.
- **Proxmox VE:** Proxmox is an open-source virtualization platform that provides easy management of virtual machines (VMs) and containers. By leveraging Proxmox, organizations can virtualize multiple data storage servers, enhancing scalability and redundancy while keeping operational complexity low.
- **ZFS (Zettabyte File System):** ZFS is a robust file system that combines disk management and volume management into a single platform. Its self-healing capabilities, high storage capacity, and snapshot functionalities make it a preferred choice for managing large amounts of data in self-hosted environments.
- **RAID Configurations:** Redundant Array of Independent Disks (RAID) setups allow data to be distributed across multiple drives to enhance performance, redundancy, and fault tolerance. Common configurations include RAID 1 (mirroring for redundancy) and RAID 5 (striping with parity for fault tolerance).

4.4.2.3 Challenges and Considerations

- **Initial Setup Costs:** Implementing self-hosted systems requires upfront investments in hardware, networking, and possibly physical space, which can be a barrier for smaller organizations.
- **Maintenance:** Self-hosted systems demand regular maintenance, including hardware checks, software updates, and security monitoring, which can require dedicated personnel or teams.
- **Scalability:** While self-hosting provides control, scaling storage capacities might be more cumbersome compared to public cloud solutions, which allow easy adjustments to storage needs.

Despite these challenges, advancements in automation and virtualization technologies make self-hosting more accessible than ever before. Self-hosted systems, when properly managed, offer a reliable and scalable alternative to public cloud solutions, ensuring data security and compliance with organizational policies.

4.5 Practical Exercises: Object-Attribute Relationships and ERDs

In this section, students will work through a series of exercises designed to develop an understanding of object-attribute relationships, create compact data representations, visualize those representations in the form of Entity-Relationship Diagrams (ERDs), and compare them to naive, table-based data representations. By the end of this section, students will have a deeper understanding of how to efficiently model data relationships and visualize them in a structured way.

4.5.1 Exercise 1: Understanding Object-Attribute Relationships

Objective: Understand the relationships between objects and their attributes.

Given is a data description in form of a university campus with a department, faculties, students and courses. Implement the following tasks:

- Formalize for each of the objects at least three attributes (e.g., courses have a title, a number, a description).
- Define at least three instances per object (e.g., the data modeling course and the machine learning course).
- Define relationships between these objects (e.g., students can register for courses, courses are organized by faculties).

Answer the following questions:

- What is the difference between the model definition and the instantiation?
- What happens if additional object categories are added?

4.5.2 Exercise 2: Visualizing Relationships with ERDs

Objective: Learn to visualize and interpret object relationships using ERDs.

Implement the following tasks:

- Implement your definitions using the pre-implemented code that is based on the *erdantic* and *pydantic* Python libraries.
- Visualize your object-attribute relationships.
- Describe all elements of the visualization.

Answer the following questions:

- The relationship between the objects is depicted by arrows with different symbols. Why are different symbols needed in ERDs?
- What happens to the visualization if new objects are added?

4.5.3 Exercise 3: Developing Compact Data Representations

Objective: Identify strategies to optimize and simplify the data representation. Implement the following tasks:

- Revisit your initial object definitions and consider how attributes like instructor and courses could be further simplified.
- Update your models to reflect these changes.
- Compare the resulting ERD to the original one.

Answer the following questions:

- What is the major difference of the two ERDs?
- What happens to the ERDs if new objects are added?

4.5.4 Exercise 4: Comparing ERDs to Table-Based Representations

Objective: Compare structured ERDs to table-based data representations. Implement the following tasks:

- Create a naive table-based model using the pre-implemented Python dictionaries or dataframes to represent the data.
- Measure the memory requirements using the pre-defined tools.
- Compare the memory requirements to the compact data representation.

Report the results and answer the following questions:

- Why are the memory requirements different (think of data redundancies)?
- How would the comparison look like when 100.000 students would be enrolled in 1.000 courses hosted by 100 faculties in 1 department?

Fig. 4.1 For the latest online version of the notebook, scan the QR code or visit: https://github.com/mul-cps/DMBook/tree/main/ex_data

4.6 Jupyter NB Exercise—Data

In this exercise, you will work through practical examples to develop essential data analysis skills. You will learn how to identify and visualize outliers, compute and interpret feature correlations, and prepare representative training and test datasets. These techniques form the foundation for meaningful data exploration and preprocessing prior to building data-driven models (Fig. 4.1).

Chapter 5
Information Theory

This chapter introduces core concepts from information theory, such as *surprise* and *entropy*. These ideas are essential for understanding data efficiency, compression, and information loss. In addition, the chapter introduces data discretization as a key technique for transforming continuous data into discrete representations, as well as the Shannon–Nyquist theorem for signal sampling.

5.1 Surprise and Shannon Information

The term *surprise* refers to something that is unexpected or new to us. For example, you are only *slightly* surprised if you flip a fair coin and it lands on *heads*. In contrast, you would be extremely surprised if you win a lottery with '6 out of 45'.

In information theory, such intuitive notions of surprise can be described mathematically. Starting with the basics, a single bit can take on either the value 0 or 1, representing two possible outcomes of an event.

With two bits, already four *possibilities* can be represented, i.e., $\{00, 01, 10, 11\}$. For three bits, eight possibilities exist, and for four bits, 16 possibilities can be modeled. In general, the number of representable possibilities is given by

$$\text{number of possibilities} = 2^n,$$

where n denotes the number of available bits.

Conversely, if one wants to compute the number of required bits to encode x possible outcomes, this is given by

$$n = \log_2(x).$$

Using these definitions, the notion of *surprise* can be quantified. This is best illustrated using the simple example of tossing multiple coins and observing the

E. Rueckert, *Data Modeling*, https://doi.org/10.1007/978-981-95-7592-3_5

same outcome each time, e.g., always *heads*. How surprised are you when

- you toss a single coin and it lands on heads? **Answer:** 'A bit', or 'one out of two possibilities'.
- you toss two coins and both land on heads? **Answer:** 'One out of four possibilities'.
- you toss four coins and all land on heads? **Answer:** 'One out of sixteen possibilities'.

Clearly, the less likely an outcome is, the more surprised we are. This intuition can be expressed mathematically by defining the information-theoretic quantity **surprise** (measured in bits) as

$$s = \log_2(x) \quad \text{[bits]},$$

where x denotes the number of possible outcomes (e.g., 2, 4, or 16 in the examples above).

5.1.1 Shannon Information

The concept of *Shannon Information*, named after **Claude Elwood Shannon** (April 30, 1916–February 24, 2001), a mathematician, electrical engineer, and computer scientist, formalizes this idea of surprise. It is defined as

$$I(X) = \log_2\left(\frac{1}{p(X)}\right), \tag{5.1}$$

where X denotes an event and $p(X)$ its probability. In this formulation, Shannon explicitly uses the inverse probability $1/p(X)$ to capture the **concept of surprise**.

For the three coin-toss examples above, the Shannon information evaluates to $\log_2(1/\{1/2\}) = 1$, $\log_2(1/\{1/4\}) = 2$, and $\log_2(1/\{1/16\}) = 4$.

Thus, rarer events yield larger information values. In other words, we are more surprised by events with low probability.

Definition 5.1 The Shannon Information I(X) expresses 'surprise' in the unit bits.

5.2 Entropy

Now consider tossing a coin every day and playing the lottery every day. After one year, which outcomes would surprise you more: the coin tosses or the lottery results?

Answering this question requires computing the **average Shannon information**, which is referred to as **Shannon entropy**.

Entropy measures the uncertainty or unpredictability of a random variable. In a dataset, it represents the minimum average number of bits required to encode the data without losing information. For example, a dataset with a balanced 50/50 class distribution has higher entropy than one in which 90% of the samples belong to a single class.

The entropy $H(X)$ of a random variable X is defined as

$$H(X) = -\sum_{i=1}^{n} p(x_i) \log_2 p(x_i),$$

where $p(x_i)$ denotes the probability of outcome x_i. A higher entropy indicates greater uncertainty and, consequently, a higher information content.

For the coin-flipping experiment, the entropy per day is

$$H(\text{Coin}) = \tfrac{1}{2} \log_2(1/\{1/2\}) + \tfrac{1}{2} \log_2(1/\{1/2\}) = 1 \text{ bit}.$$

For the lottery, where the probability of winning '6 out of 45' is approximated by 2^{-23}, the entropy per day is

$$\begin{aligned} H(\text{Lottery}) &= \tfrac{1}{2^{23}} \log_2(1/\{1/2^{23}\}) + \left(1 - \tfrac{1}{2^{23}}\right) \log_2\left(\frac{1}{1 - 1/2^{23}}\right) \\ &= 0.00000292 \text{ bits}. \end{aligned}$$

The key observation is that, *on average, we are most surprised when all possible events are equally likely.*

Definition 5.2 The Shannon entropy computes the sum of the information of all possibilities, weighted by their probability in the unit bits.

5.2.1 Optimal Encoding

The Shannon entropy provides a theoretical lower bound on the most efficient encoding of symbols. To illustrate this, consider transmitting letters composed of only four symbols: $\{A, B, C, D\}$. Each symbol is encoded using bits.

For a **fixed-length encoding**, two bits are required to represent all four letters. One possible encoding is:

$$A \to 00, \quad B \to 01, \quad C \to 10, \quad D \to 11.$$

If all symbols are equally likely, this encoding is optimal. However, if the symbols have different probabilities, the problem becomes an interesting *optimal encoding* challenge. Assume the following probabilities:

$$p(A) = \frac{1}{3}, \quad p(B) = \frac{1}{2}, \quad p(C) = p(D) = \frac{1}{12}.$$

To quantify encoding efficiency, we define the **Average Bits per Symbol (ABS)**:

$$ABS(\text{Encoding}) = \sum_{x \in X} p(x) L_x, \tag{5.2}$$

where L_x is the number of bits used to encode symbol $x \in X$.
For the fixed-length encoding above, we have:

$$ABS(\text{fixed-length}) = 2 \text{ bits per symbol.}$$

In contrast, for a **variable-length encoding**, such as

$$A \to 11, \quad B \to 0, \quad C \to 100, \quad D \to 101,$$

the ABS measure becomes

$$\begin{aligned} ABS(\text{variable-length}) &= \sum_{x \in X} p(x) L_x \\ &= \frac{1}{3} \cdot 2 + \frac{1}{2} \cdot 1 + \frac{1}{12} \cdot 3 + \frac{1}{12} \cdot 3 \\ &= 1.67 \text{ bits per symbol.} \end{aligned}$$

Finally, the **optimal encoding** is determined using the Shannon entropy:

$$\begin{aligned} H(X) &= - \sum_{x \in X} p(x) \log_2(p(x)) \\ &= - \left\{ \frac{1}{3} \log_2 \frac{1}{3} + \frac{1}{2} \log_2 \frac{1}{2} + 2 \cdot \frac{1}{12} \log_2 \frac{1}{12} \right\} \\ &= 1.63 \text{ bits per symbol.} \end{aligned}$$

Comparing the three measures (2 bits, 1.67 bits, and 1.63 bits) shows that the chosen variable-length encoding is already quite close to the theoretical optimum. While for simple examples with only four symbols, manually defining encoding schemes is feasible, Shannon entropy provides a systematic tool to optimize encoding in more complex, real-world scenarios.

Definition 5.3 The Shannon entropy gives the theoretical lower bound on the average bits per symbol required to encode symbols. It provides a benchmark for the best possible compression of data.

5.3 Data Discretization

Data discretization is the process of transforming continuous data or variables into discrete categories, facilitating certain types of analysis and machine learning algorithms. This transformation is essential in scenarios where data interpretation, storage, or model simplification is required. Discretization is particularly valuable in decision tree algorithms, Naive Bayes classifiers, and any context where simplifying the input space improves performance or interpretability.

Techniques for Data Discretization:

- **Equal-width Binning:** The range of values is divided into equal-width intervals. For example, if the range of a variable is from 0 to 100, you can divide it into 10 intervals of equal width (e.g., 0–10, 11–20, and so on). This method is simple to implement, but can sometimes produce suboptimal results if the data distribution is skewed.
- **Equal-frequency Binning:** In this approach, the data is divided into intervals such that each bin contains approximately the same number of instances. This method can be more effective when dealing with skewed data, as it adjusts the binning to reflect the actual data distribution.
- **Cluster-based Discretization:** Techniques like K-means clustering can be used to group continuous data into discrete clusters based on similarity. The resulting clusters represent discretized values and capture patterns in the data effectively.
- **Decision Tree-based Discretization:** A decision tree algorithm can be used to partition data into discrete intervals based on the values that lead to the highest information gain, which is highly effective in minimizing entropy.
- **Entropy-based Discretization:** Leveraging information theory, data can be split into intervals that reduce the overall entropy. This technique ensures that each interval is as homogenous as possible concerning the target class in supervised learning tasks.

5.3.1 Example of Entropy-Based Discretization

Suppose you have a dataset with continuous age values and a target variable for disease presence (yes/no). Using entropy-based discretization, you might find that splitting the ages into intervals like 0−20, 21−40, and 41−60 optimizes the predictive accuracy for disease detection by reducing the entropy in each interval.

Remark 5.1 Discretization must be done carefully, as it can affect model performance and data interpretation. While it simplifies data, over-discretization can result in information loss.

5.4 Shannon-Nyquist Theorem

In addition to these methods, the Shannon-Nyquist theorem plays a role in signal discretization, particularly in digital signal processing. This theorem states that a continuous signal can be completely represented by discrete samples if the sampling rate is at least twice the highest frequency present in the signal.

The Shannon-Nyquist Theorem not only sets the minimum sampling rate for accurate signal conversion, but it also provides a foundation for signal reconstruction. Once the signal is sampled at a rate greater than or equal to the Nyquist rate, the original continuous signal can be reconstructed from these discrete samples using the reconstruction formula.

5.4.1 Analog Signal Reconstruction

The reconstruction is typically done using an ideal low-pass filter to interpolate between the samples. The formula for the reconstructed signal $x(t)$ is:

$$x(t) = \sum_{n=-\infty}^{\infty} x_n \operatorname{sinc}\left(\frac{t - nT}{T}\right),$$

where x_n are the discrete samples. The symbol T denotes the sampling interval ($T = 1/f_s$) and $\operatorname{sinc}(t) = \sin(\pi t)/(\pi t)$ is the sinc function, which acts as the interpolation function.

This formula ensures that the continuous signal is reconstructed by summing all the sinc functions, each centered on a sample point and weighted by the sample's amplitude.

In simpler terms, each sample contributes a sinc-shaped pulse, and the sum of all these pulses perfectly reconstructs the original signal if the sampling rate meets the Nyquist criterion.

This process prevents aliasing and allows for an accurate reconstruction of the original signal from its sampled points.

5.5 Practical Exercises: Entropy Calculation and Signal Discretization

Through these exercises, students will gain practical experience in computing entropy and applying the Shannon-Nyquist theorem to analog signals. The exercises aim to deepen understanding of information theory concepts, linking them to real-world data discretization tasks and their implications in various applications, such as data compression and signal processing.

5.5.1 Exercise 1: Entropy Calculation

Objective: Compute the entropy of a simple sentence using a limited set of symbols.
Implement and test the following tasks:

- Choose a Sentence: Select a simple sentence, for example: 'Data science is fascinating.'
- Define a Reduced Set of Symbols: Use a limited character set that includes only the letters and some symbols, e.g., {a, d, e, f, g, i, n, o, r, s, t, c, h, space}.
- Calculate Symbol Frequencies, where you count the occurrences of each symbol in the chosen sentence.
- Compute the Entropy by using the entropy formula: $H(X) = -\sum_{i=1}^{n} p(x_i) \log_2 p(x_i)$, where $p(x_i)$ is the frequency of each symbol divided by the total number of symbols.

Report the results. Answer the following questions:

- How can one compare the entropy of two sentences? What is the underlying assumption?
- How can the entropy results be used to compress the sentences?
- Explore how varying the reduced set of symbols impacts the computed entropy. What patterns do you observe?

5.5.2 Exercise 2: Signal Discretization and Reconstruction

Objective: Compute the lowest discretization frequency for a given analog signal, discretize it, and reconstruct it.

Given is an analog signal generator. Discretize the analog signal by executing the following processes:

- Visualize the Analog Signal: Start by inspecting a predefined analog signal, such as a sine wave. Plot the signal using Python to understand its characteristics.
- Determine the Maximum Frequency: Compute the maximum frequency present in the signal (hint: $f = 1/T$).
- Calculate the Lowest Discretization Frequency: Based on the Shannon-Nyquist theorem, calculate the lowest sampling frequency: $f_s = 2f$, where f is the highest frequency in the signal.
- Discretize the Analog Signal: Sample the signal at the calculated frequency and store the discrete values in an array.
- Visualize the Analog and Discrete Representations: Create plots to compare the original analog signal and its discretized version. Discuss what happens if the sampling frequency is significantly increased or decreased.
- Reconstruct the Analog Signal: Use the sinc function for reconstruction: $x(t) = \sum_{n=-\infty}^{\infty} x_n, \operatorname{sinc}\left((t - nT)/T\right)$, where $T = 1/f_s$.

- Compare the Reconstruction with the Original Signal: Analyze the accuracy of the reconstruction visually and compute a similarity measure (e.g., Mean Squared Error) between the original and reconstructed signals.

  Report the results. Answer the following questions:

- How does the sampling frequency affect the reconstruction error?
- What is the drawback of using extremely high sampling frequencies?
- What happens in real world applications when the sampling frequency is to low, e.g., it is limited by the hardware used?

5.5.3 *Exercise 3: Entropy-Based Signal Discretization*

Objective: Define and test a conceptual rule that utilizes data entropy to improve signal discretization.

Define your own rule how the entropy can shape your discretization process. For example, the entropy can be used to adjust the sampling frequency based on the signal characteristics. Test the following processes:

- Apply your rule by adjusting the pre-implemented algorithm.
- Compare the results with the original approach.
- Visualize the results using metrics like reconstruction accuracy and computational efficiency.

  Report the results. Answer the following questions:

- Did your entropy-based rule improve the discretization process?
- What are the assumptions of your rule?
- What are the limitations of your rule?
- Would your rule generalize to any analog signals?

5.6 Jupyter NB Exercise—Information Theory

In this exercise, you will explore fundamental concepts of information theory, with a focus on entropy and the discretization of analog signals. You will begin by computing the entropy of self-selected sentences to develop intuition for measuring information content in text.

Next, you will examine how continuous signals can be represented in digital form. Using a sine wave as an example, you will perform sampling to convert an analog signal into a discrete sequence and then reconstruct the original waveform from the sampled data. This hands-on process illustrates the core principles connecting the analog and digital domains in modern communication and signal processing systems.

Fig. 5.1 For the latest online version of the notebook, scan the QR code or visit: https://github.com/mul-cps/DMBook/tree/main/ex_informationtheory

Finally, you will implement and test an entropy-based signal discretization method, reinforcing the relationship between information content, sampling, and data representation (Fig. 5.1).

Chapter 6
Analyses

In this chapter, we explore various data analysis strategies, particularly elementary concepts from applied statistics and data visualization strategies, which form the foundation for data modeling.

6.1 Applied Statistics

Statistical approaches are essential to describe and compare datasets.

6.1.1 Basic Statistics

The basic statistics include measures of central tendency, such as:

- **Mean:** Given a dataset $\mathcal{D} = \{x_1, x_2, \ldots, x_N\}$, the mean (or average) is computed as:

$$\mu = \frac{1}{N}\sum_{i=1}^{N} x_i .$$

- **Median:** The value separating the higher half from the lower half of the data. For a dataset sorted in ascending order, if n is odd, the median is $x_{\frac{N+1}{2}}$, and if N is even, the median is the average of the two middle values.
- **Mode:** The value(s) that appear most frequently in the dataset.

E. Rueckert, *Data Modeling*, https://doi.org/10.1007/978-981-95-7592-3_6

Measures of dispersion (variability in the data) include:

- **Variance:** Quantifies the spread of the data around the mean and is given by:

$$\sigma^2 = \frac{1}{N}\sum_{i=1}^{N}(x_i - \mu)^2.$$

- **Standard Deviation:** The square root of variance, representing the average distance from the mean:

$$\sigma = \sqrt{\frac{1}{N}\sum_{i=1}^{N}(x_i - \mu)^2}.$$

- **Range:** The difference between the maximum and minimum values in the dataset:

$$\text{Range} = x_{\text{max}} - x_{\text{min}}.$$

Remark 6.1 Assuming normally distributed data is a frequently underestimated assumption. The use of mean and standard deviation as statistical representations of data assumes that the data follows a normal distribution (or Gaussian distribution). This assumption is crucial because in a normally distributed dataset, the mean serves as a reliable measure of central tendency, and the standard deviation effectively captures the data's spread.

In such distributions, about 68% of the data falls within one standard deviation of the mean, 95% within two standard deviations, and 99.7% within three standard deviations, making these metrics robust descriptors. However, when the data is skewed or follows a non-Gaussian distribution, the mean can become less informative, as it may not accurately represent the dataset's center, especially in the presence of outliers. Similarly, the standard deviation may not fully capture the spread or dispersion (Fig. 6.1).

For example, in highly skewed data (e.g., income distributions), the median or interquartile range might provide a better understanding of the data's center and variability. Hence, when using mean and standard deviation, it's essential to first validate whether the data approximates a normal distribution through techniques like visual inspection (e.g., histograms) or statistical tests (e.g., the Shapiro-Wilk test). If the assumption doesn't hold, alternative descriptive statistics or transformations of the data should be considered.

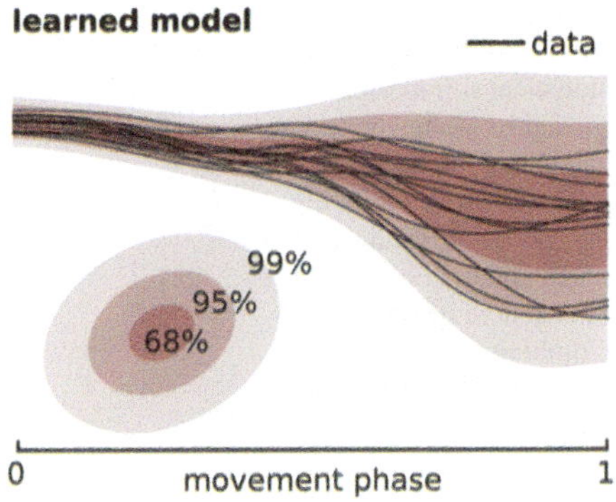

Fig. 6.1 This figure from [26] illustrates a Gaussian model that represents trajectories denoted by black lines. The shaded areas represent the data distributions that fall within one standard deviation of the mean (68%), within two standard deviations (95%), and within three standard deviations (99.7%)

6.2 Data Visualization

A powerful and common practice for analyzing data is the visual inspection. Statistics or projections to lower-dimensional spaces, e.g., through dimensionality reduction techniques and clustering algorithms, can be utilized.

6.2.1 Data Distribution Plots

Data distribution plots visually represent the distribution of data points. A commonly used plot is the box plot, which shows the minimum, first quartile, median, third quartile, and maximum values. The density plot is another tool used to estimate the probability distribution of a dataset, giving a smoother alternative to histograms.

Whenever possible, statistics should always be presented jointly with the raw data to identify any loss of information due to the abstraction. An example is shown in Fig. 6.2. This figure illustrates the movement onset times of human subjects performing goal-directed tasks [26].

6.2.2 Histograms

A histogram divides the range of data into bins and counts the number of observations in each bin. Mathematically, it is a function:

$$f(x) = \frac{\text{Count of values in bin } B_i}{N \cdot \text{width of } B_i},$$

where N is the number of data points, and B_i is the i-th bin.

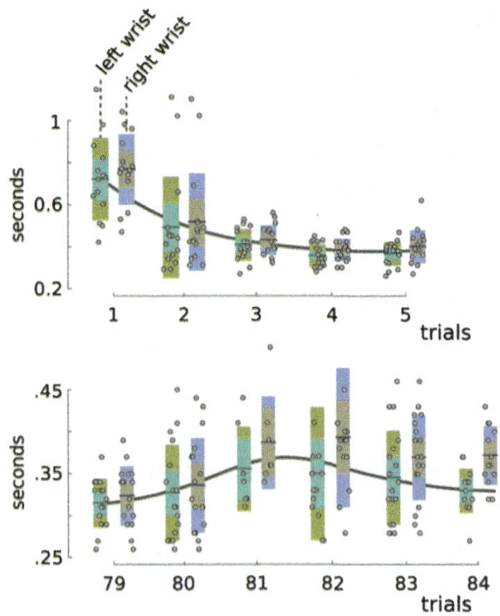

Fig. 6.2 This figure from [26] illustrates how statistical summaries can be presented alongside raw data. The dots represent individual measurements, while the horizontal bars indicate the mean and the colored error bars denote quantiles. Combining raw data and statistical summaries in a single visualization helps to assess potential information loss and to evaluate whether assumptions—such as Gaussianity used for computing the mean—are valid

6.2.3 Covariance and Correlation Plots

The covariance between two variables X and Y is a measure of how much they change together and is calculated as:

$$\mathrm{Cov}(X, Y) = \frac{1}{N}\sum_{i=1}^{N}(x_i - \mu_X)(y_i - \mu_Y).$$

The correlation coefficient (Pearson's correlation) normalizes the covariance and is given by:

$$\rho(X, Y) = \frac{\mathrm{Cov}(X, Y)}{\sigma_X \sigma_Y}.$$

Correlation plots can visually represent how strongly two variables are related. An example is shown in Fig. 6.3.

6.2.4 Data Clustering

Clustering is a technique used to group data points based on similarity. The goal is to assign each point to a cluster such that points in the same cluster are more similar to each other than to those in other clusters.

Fig. 6.3 This figure shows how correlations between multiple features can be visualized. Strong positive correlations are indicated by values close to 1.0, while values near 0 indicate weak or no correlation

6.2.4.1 Dimension Reduction Techniques

A common method for managing high-dimensional datasets is dimension reduction, which simplifies the data by reducing the number of variables while retaining essential information. One widely used technique is Principal Component Analysis (PCA) [3], which transforms the data into a new coordinate system that maximizes variance of the transformed data and identifies the most important patterns.

The transformation can be expressed as:

$$\boldsymbol{Y} = \boldsymbol{X}\boldsymbol{W},$$

where:

- $\boldsymbol{X}$ is the original data matrix,
- $\boldsymbol{W}$ is a matrix of eigenvectors (principal components), and
- $\boldsymbol{Y}$ is the transformed lower-dimensional data.

Each principal component in $\boldsymbol{Y}$ represents a linear combination of the original variables, arranged in order of their explained variance or 'importance'. By selecting a subset of the principal components (e.g., the first two components), we can reduce the dimensionality while still preserving much of the data's variability. This reduced representation allows for visualization in 2D plots, making it easier to uncover underlying patterns or clusters in the data.

By focusing on the most significant components, techniques like PCA enable both visualization and the application of machine learning algorithms to otherwise complex, high-dimensional datasets in a more computationally efficient manner. The drawback of applying this dimensionality reduction technique is the loss of information.

6.3 Normalization and Standardization

When working with data, it's crucial to ensure that the features are scaled properly, particularly when dealing with machine learning algorithms that are sensitive to the magnitudes of the inputs, such as gradient-based optimization algorithms or distance-based algorithms like k-means clustering. Two common techniques used for scaling data are **Normalization** and **Standardization**.

6.3.1 Normalization (Min–Max Scaling)

Normalization, also known as **Min–Max Scaling**, adjusts the values in a dataset to fall within a specific range, typically between 0 and 1. This method is particularly useful when the data is not normally distributed or exhibits significant variation in scale. Normalization is especially effective for algorithms that do not assume a Gaussian distribution, and its appropriateness can be evaluated by visualizing the data using histograms or conducting tests for normality.

Normalization preserves the relative relationships between data points. The formula for normalization is:

$$x' = \frac{x - x_{\min}}{x_{\max} - x_{\min}},$$

where:

- x is the original data point,
- $x_{\min}$ and $x_{\max}$ are the minimum and maximum values in the dataset, respectively, and
- x' is the normalized value of x.

By applying this transformation, all the data points are scaled to the [0, 1] range, where the minimum value becomes 0 and the maximum value becomes 1. This technique is commonly used when the features are on different scales or units, and the model you are using expects the data to be bounded.

Remark 6.2 Normalization can be sensitive to outliers because extreme values can significantly affect the scaling. Therefore, it is important to identify and to remove outlier before applying normalizing the data.

6.3.2 Standardization (Z-Score Normalization)

Standardization, also called **Z-score Normalization**, transforms the data to have a mean of 0 and a standard deviation of 1. Unlike normalization, standardization does not bound the values to a specific range but rather scales the data based on its distribution.

The formula for standardization is:

$$x' = \frac{x - \mu}{\sigma},$$

where:

- x is the original data point,
- μ is the mean of the dataset,
- σ is the standard deviation of the dataset, and
- x' is the standardized value of x.

This transformation centers the data around zero and rescales it based on the data's standard deviation. Standardization is commonly used when the data follows a Gaussian (normal) distribution or when features have different units or scales but the same importance.

Remark 6.3 Standardization assumes that the data follows a Gaussian (normal) distribution, which needs to be ensured before applying this technique. Standardization is less sensitive to outliers compared to *Normalization*, because it focuses on the distribution of the data.

6.3.3 Choosing Between Normalization and Standardization

The choice between normalization and standardization depends on the specific context and the algorithms being used:

- **Normalization** is typically used when you want to maintain relative relationships between data points, or when the data does not follow a Gaussian distribution.
- **Standardization** is often more appropriate when the data is normally distributed or when you're working with algorithms that assume Gaussian distributions (e.g., linear regression, logistic regression) or rely on distances (e.g., k-means, SVM).

In practice, many machine learning pipelines automatically apply one of these techniques, so it's important to understand the data's distribution and the algorithm's requirements before deciding which scaling method to apply.

6.4 Alternative Data Scaling Techniques

In addition to normalization and standardization, there are several alternative data scaling techniques that can be applied depending on the characteristics of the dataset and the machine learning algorithm being used. Each method addresses different issues such as outliers, non-Gaussian distributions, or the need to preserve specific data structures. Below, some widely-used alternative scaling techniques are discussed.

6.4.1 *Robust Scaler*

The **Robust Scaler** is designed to mitigate the effects of outliers by using statistics that are less sensitive to extreme values, such as the median and the interquartile range (IQR). The scaling transformation is given by:

$$x' = \frac{x - \text{median}(x)}{\text{IQR}(x)},$$

where:

- $\text{median}(x)$ is the median of the data, and
- $\text{IQR}(x)$ is the interquartile range, defined as the difference between the 75th and 25th percentiles: $\text{IQR}(x) = Q_3(x) - Q_1(x)$.

This method is useful when the data contains outliers that could skew the results of other scaling techniques like min-max scaling.

6.4.2 *Max Abs Scaling*

The **Max Abs Approach** scales data by dividing each value by the maximum absolute value in the dataset, ensuring that the data is scaled within the range of -1 to 1 without changing the data's sparsity.

$$x' = \frac{x}{\max(|x|)}.$$

This scaling is particularly useful for sparse datasets and algorithms that require non-negative inputs, such as text data represented in sparse matrices.

6.4.3 *Logarithmic Transformation*

A **Logarithmic Transformation** is often used when the data contains values that span several orders of magnitude or exhibit an exponential growth pattern. The logarithmic transformation compresses the range of values, reducing the impact of large outliers. The transformation is given by $x' = \log(x + 1)$.

This transformation is particularly useful for skewed data, such as financial or population data, where small values are common, but large values dominate the distribution.

6.4.4 Power Transformation (Box-Cox and Yeo-Johnson)

Power transformations are used to stabilize variance and make data more normally distributed. Two popular power transformations are Box-Cox and Yeo-Johnson. Below, the box-cox transformation is described and illustrated.

The Box-Cox transformation is for positive data only. It is defined by

$$x' = \frac{x^{\lambda} - 1}{\lambda}, \quad \lambda \neq 0 .$$

If $\lambda = 0$, the transformation becomes a logarithmic transformation: $x' = \log(x)$. Note that the the approach can only be used for positive values, i.e., $x > 0$, and λ has to be adapted to the individual datasets.

Figure 6.4 illustrates the effect of the Box–Cox transformation on three example distributions, where the hyper-parameter λ is chosen via a maximum likelihood approach for each of the three distributions.

6.4.5 Mean Normalization

Mean Normalization rescales the data so that its mean becomes 0, and it typically scales values to a range of -1 to 1. The formula is:

$$x' = \frac{x - \mu}{x_{\max} - x_{\min}} .$$

This method is often used when you want to center the data and scale it within a specific range.

Remark 6.4 Each scaling method has its own strengths and is suited for different scenarios. For example, if outliers are present, Robust Scaler or Logarithmic Transformation may be more appropriate than Min-Max scaling. When using algorithms sensitive to distance metrics, such as k-means clustering, L2 normalization can help standardize the data. The choice of scaling method should be guided by the data's characteristics and the requirements of the model being used.

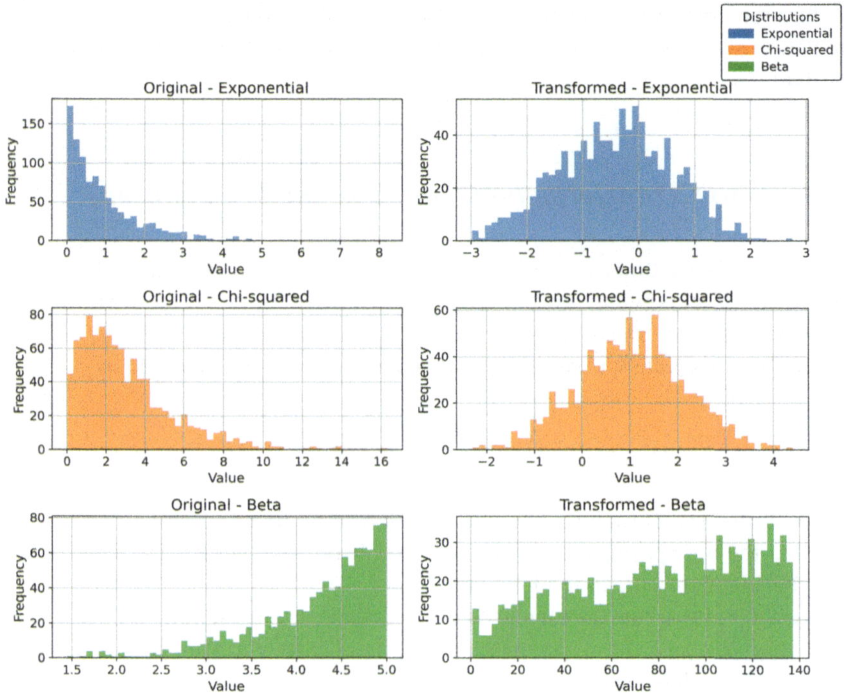

Fig. 6.4 This figure illustrates the effect of the Box–Cox transformation on three example distributions (Exponential, Chi-squared, and a Beta). For each dataset, the left panel shows the original histogram and the right panel shows the transformed data. The transformation reduces skewness, stabilizes variance, and brings the data closer to normality

6.5 Data Issues

An important pre-processing step is to detect and resolve data issues.

6.5.1 Outlier Detection Strategies

Outliers are data points significantly different from others in a dataset. Some common detection methods include:

The Z-score method measures how many standard deviations a data point is from the mean:

$$Z = \frac{x - \mu}{\sigma} ,$$

where data points with $\|Z\| > 3$ are considered as outliers.

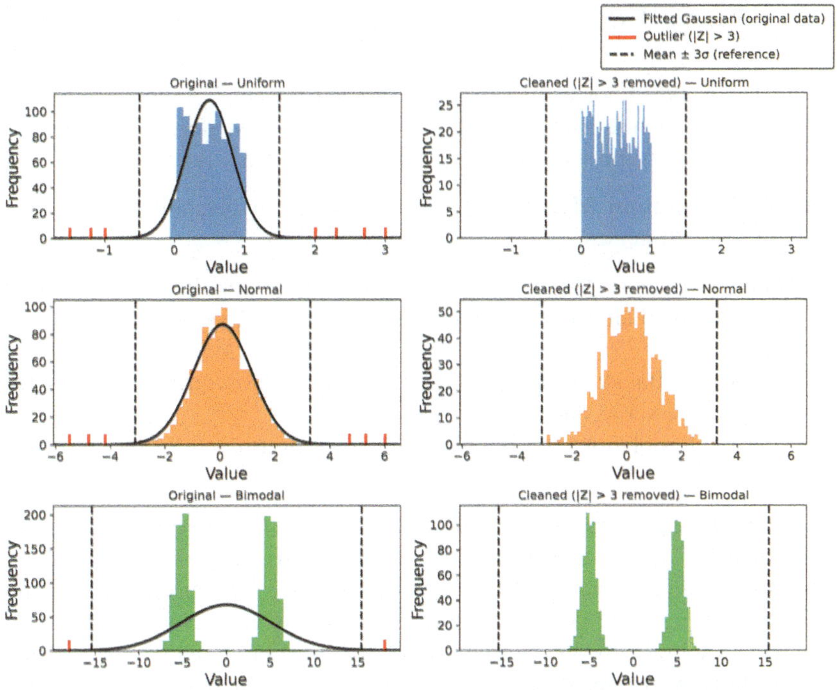

Fig. 6.5 Z-score outlier removal with three example datasets (Uniform, Normal, and bimodal). The left column shows the original histograms with outliers marked and fitted Gaussian curves based on the sample mean and standard deviation; dashed lines indicate $\mu \pm 3\sigma$. The right column shows the cleaned datasets after removing the flagged outliers. The bimodal example illustrates a limitation: global Z-scores can under- or mis-flag outliers in multimodal data due to the inflated overall standard deviation

Figure 6.5 shows three datasets—Uniform, Normal, and bimodal—processed with Z-score outlier removal. The left column displays the original histograms with fitted Gaussian curves and reference lines at $\mu \pm 3\sigma$, with outliers marked; the right column shows the cleaned data. While the method performs as expected for approximately normal data, the bimodal example highlights a key limitation: global Z-scores can be misleading for non-normal or multimodal distributions.

Interquartile Range (IQR) Method

A complementary, robust alternative is the Interquartile Range (IQR) method. Let Q_1 and Q_3 denote the first and third quartiles. Define and apply the outlier rule as

$$\text{IQR} = Q_3 - Q_1,$$
$$x < Q_1 - 1.5\,\text{IQR} \quad \text{or} \quad x > Q_3 + 1.5\,\text{IQR}.$$

This rule is less sensitive to non-normality and extreme values than Z-scores. For stricter outlier removal strategies, a larger multiplier, like 3 IQR can be used.

6.5.2 *Missing Values*

Missing data is a common challenge in real-world datasets. Ignoring missing values can lead to bias, while using inappropriate imputation strategies may distort the dataset. Below are some standard approaches to handle missing values.

- **Simple Imputation Methods:** These techniques involve filling missing data using simple strategies such as:
  - Mean, Median, or Mode: Replace missing values with the mean (for continuous data), median (to avoid the influence of outliers), or mode (for categorical data).
  - Forward or Backward Filling: Use previous or subsequent values to fill the gaps in time-series data.
  - Linear Interpolation: Applies linear interpolation based on two neighboring values.
- **K-Nearest Neighbors (KNN):** This technique imputes missing values based on the values of the K nearest neighbors. KNN finds the most similar instances (based on distance metrics like Euclidean distance) and uses their average or majority values to fill in the missing data. This is effective when the missing values are not random and similar observations can be found within the dataset.
- **Multiple Imputation**: In this more sophisticated technique, missing values are estimated multiple times to capture the uncertainty about what the true value might have been. The multiple datasets are then used to train models, providing more robust predictions.
- **Predictive Imputation**: Machine learning models such as regression or classification can be trained to predict missing values based on the rest of the data.

6.6 Data Similarity Analyses

In data modeling, a common task is compare datasets or models. Two principled approaches can be applied.

6.6.1 *Parametric Approaches*

Parametric similarity methods assume the data follows a specific distribution, such as a Gaussian distribution. To compute the similarity of two Gaussian distributions, we use the notation $\mathcal{N}_1(\mu_1, \sigma_1)$ and $\mathcal{N}_1(\mu_2, \sigma_2)$, with the mean μ and the variance σ.

- **t-tests** are used to verify the hypothesis that the means of two groups are equal. The formula for the t-statistic is:

$$t = \frac{\mu_1 - \mu_2}{\sqrt{\frac{\sigma_1^2}{N_1} + \frac{\sigma_2^2}{N_2}}},$$

  where the number of elements used to compute the statistics are denoted by N_1 and N_2.
- **Kullback-Leibler Divergence for Univariate Gaussian Distributions** is another parametric approach that evaluates the similarity of two Gaussian models by:

$$KL\,(\mathcal{N}_1 \parallel \mathcal{N}_2) = \log\left(\frac{\sigma_2}{\sigma_1}\right) + \frac{\sigma_1^2 + (\mu_1 - \mu_2)^2}{2\sigma_2^2} - \frac{1}{2}.$$

- **Kullback-Leibler Divergence for Multivariate Gaussian Distributions** can be applied for Gaussian distributions over vectors. Thus, the mean is denoted by the vector notation $\boldsymbol{\mu}$ and $\boldsymbol{\Sigma}$ denotes the covariance matrix. The KL divergence is given by:

$$KL(\mathcal{N}_1 \parallel \mathcal{N}_2) = \frac{1}{2}\log\frac{|\boldsymbol{\Sigma}_2|}{|\boldsymbol{\Sigma}_1|} - d + \mathrm{tr}(\boldsymbol{\Sigma}_2^{-1}\boldsymbol{\Sigma}_1) + (\boldsymbol{\mu}_2 - \boldsymbol{\mu}_1)^\top \boldsymbol{\Sigma}_2^{-1}(\boldsymbol{\mu}_2 - \boldsymbol{\mu}_1),$$

  where d denotes the dimension of the vector $\boldsymbol{\mu}$. The symbol $|\boldsymbol{\Sigma}|$ denotes the determinant of the covariance. The matrix trace is computed as the sum of all diagonal elements, i.e., $\mathrm{tr}(\boldsymbol{A}) = \sum_{i=1}^{d} A_{i,i}$.

6.6.2 Non-Parametric Approaches

Non-parametric approaches do not assume any underlying distribution. One common technique is the Kullback-Leibler (KL) divergence, which measures how one probability distribution differs from a second:

$$D_{KL}(P||Q) = \sum_i P(i)\log\left(\frac{P(i)}{Q(i)}\right),$$

where P and Q are probability distributions. As an example, $P(i)$ can denote the frequency count of the distribution P for the i-th bin, see Fig. 6.6.

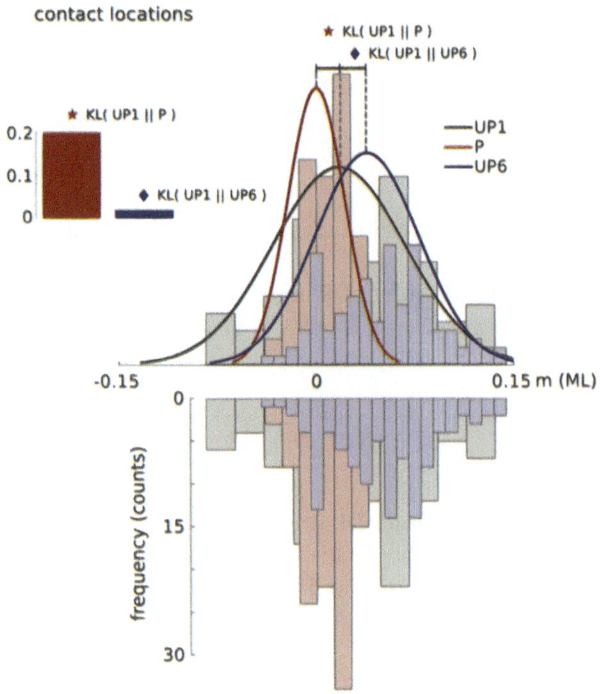

Fig. 6.6 This figure from [26] illustrates how three different distributions can be compared visually and analytically. The three distributions represent target acquisition errors in meters from three different experiments, where *UP* denotes an undisturbed environment and *P* denotes a disturbed environment. The index denotes the session number in the range from 1 to 6. The analytical comparison evaluates the similarity of two distributions using the Kullback-Leibler divergence measure (KL) under the assumption of Gaussian models

6.7 Practical Exercises: Analyzing Real-World Datasets

The following practical exercises will showcase how outliers can be detected, visualized and how feature correlations can be computed. Here are two example datasets:

- Student Performance Factors, https://www.kaggle.com/datasets/lainguyn123/student-performance-factors/data.
- Urban Air Quality and Health Impact Analysis, https://www.kaggle.com/datasets/abdullah0a/urban-air-quality-and-health-impact-dataset.

6.7.1 Exercise 1: Basic Statistics and Visualizations

Objective: Learn how to identify and visualize outliers in a real-world dataset. Given is a dataset that contains some issues. Try to identify most of them by:

- Loading the public dataset.
- Computing basic statistics (mean, variance, median).
- Detecting outliers using the Z-score method with $Z = \frac{x-\mu}{\sigma}$, where data points with $\|Z\| > 3$ are considered as outliers.

- Visualizing the data by creating histograms. Denote outliers by using a different color or symbol.

  Report the number of identified outliers. Answer the following questions:

- Is the Z-score method an adequate strategy for this dataset?
- Do the outliers have something in common?
- Without removing the outlier, would that impact the modeling results?

6.7.2 Exercise 2: Analyze Feature Correlations

Objective: Learn how to identify outliers or artifacts in the data by investing feature correlations.

Start by using the original dataset with all outliers that were detected in the previous exercise.

- Analyze feature correlations using a correlation plot.
- Test the pre-implemented k-means clustering algorithm on selected features.

  Answer the following questions:

- Do you observe any patterns?
- How can one choose k?
- Now perform the same steps using the pre-processed dataset, where outliers are removed based on the Z-score method. What do you observe?
- Adapt the Z-score method parameter and run the experiment again. How many clusters do you observe?

6.7.3 Exercise 3: Create Representative Training and Test Datasets

Objective: Understand how representative training and test dataset splits can be generated based on quantitative measures.

Implement and test the following steps:

- Create train and test datasets based on dividing the data set into two parts.
- Create histograms of the features of the two datasets within one plot per feature.
- Compare the distributions of the two datasets using the discrete Kullback-Leibler (KL) divergence.

  Answer the following questions:

- Repeat the process, but now draw random samples for the train and test splits. What do you observe?
- What happens if you decrease the number of test samples?

6.7.4 *Exercise 4: Parametric Similarity Measures of Datasets*

Objective: Compare non-parametric and parametric similarity measures on real-world datasets.

Implement and test the following steps:

- Create train and test datasets based on dividing the data set into two parts, where for testing only 10% of the data is used.
- Create histograms of the features of the two datasets within one plot per feature.
- Compare the distributions of the two datasets using the Kullback-Leibler (KL) divergence for Gaussian distributions.

Report the similarity score. Answer the following questions:

- What is the assumption when dataset similarities are computed with this KL divergence measure?
- What happens if the test dataset consists only of 2% of the total number of samples?

6.8 Jupyter NB Exercise—Analyzing Real-World Datasets

In this exercise, you will explore fundamental concepts of information theory, with a focus on entropy and the discretization of analog signals. You will begin by computing the entropy of self-selected sentences to develop intuition for measuring information content in text.

Next, you will examine how continuous signals can be represented in digital form. Using a sine wave as an example, you will perform sampling to convert an analog signal into a discrete sequence and then reconstruct the original waveform from the sampled data. This hands-on process illustrates the core principles connecting the analog and digital domains in modern communication and signal processing systems.

Finally, you will implement and test an entropy-based signal discretization method, reinforcing the relationship between information content, sampling, and data representation (Fig. 6.7).

Fig. 6.7 For the latest online version of the notebook, scan the QR code or visit: https://github.com/mul-cps/DMBook/tree/main/ex_dataanalysis

Part IV
Machine Learning Fundamentals

In this chapter, we discuss data organization strategies, including data splitting and data augmentation techniques, model evaluation techniques using k-fold cross-validation, and we present illustrative examples of selected machine learning examples applied to real-world data modeling problems.

Chapter 7
Data Organization

In this chapter, data organization strategies for preparing datasets for machine learning algorithms are discussed.

7.1 Data Splitting Strategies

In machine learning, it is crucial to evaluate the performance of models on data that they have not seen during training [21, 34]. This ensures that models generalize well to new, unseen data. To achieve this, data splitting strategies are employed to partition the available data into distinct sets: a **training set**, a **validation set**, and a **test set**.

7.1.1 Three Key Datasets

In general, machine learning tasks rely on three datasets, each serving a distinct purpose:

- **Training Dataset:** The model learns patterns and relationships from this dataset. The training process involves adjusting model weights or parameters to minimize an objective function (such as a loss function). A crucial evaluation is the dependency of the model's performance on the number of training samples. While simpler problems, like linear regression tasks, can be trained from a few samples, deep neural networks processing high-dimensional data like text or images require millions of samples.
- **Validation Dataset:** This set is used for hyperparameter tuning and to monitor the model's performance during training. The model's evaluation on this set helps prevent overfitting and ensures that the optimal configuration of hyperparameters

E. Rueckert, *Data Modeling*, https://doi.org/10.1007/978-981-95-7592-3_7

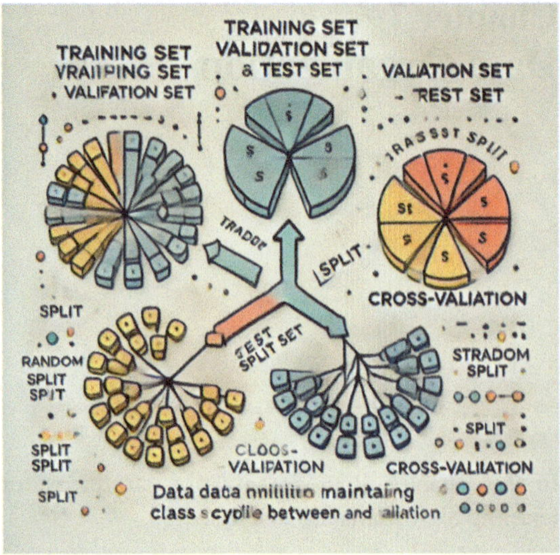

Fig. 7.1 AI-generated sketch illustrating various data splitting strategies commonly used in machine learning, such as training/test splits and cross-validation

(such as learning rate, regularization strength, etc.) is selected. It is important to note that validation data should not be used to train the model or adjust weights.

- **Test Dataset:** After the model is trained and hyperparameters have been optimized using the validation dataset, the test dataset is used to evaluate the final model's performance. This data is completely isolated from the training process, ensuring an unbiased evaluation of the model's ability to generalize to unseen data (Fig. 7.1).

Remark 7.1 It is important to note that the underlying assumption in applying data splitting strategies is that all data samples in all sets originate from the same gernative distribution. Thus we assume that all samples are *independent and identically distributed (i.i.d.)* random variables. In other words, we assume that we use different data samples per set but they all represent the same physical process and also fully cover the complete space.

7.2 Data Augmentation Strategies

Data augmentation is an approach for handling small or imbalanced datasets. The core idea behind data augmentation is to create new synthetic samples that retain the important properties, correlations and patterns from the original data without distorting its underlying structure. This technique is used to increase the diversity of the training data and can help models generalize better.

7.2.1 *Motivation for Data Augmentation*

There are a number of reasons why data augmentation can improve the overall quality and performance of a data modeling process.

- **Overfitting Prevention:** Augmentation increases the training data size, helping reduce the likelihood that the model will overfit to a small dataset.
- **Data Scarcity:** In real-world applications, acquiring large datasets is often challenging. Augmentation allows models to work with limited data.
- **Class Imbalance:** In classification tasks, some classes may be under-represented. Data augmentation can help by creating synthetic samples of the minority class.

7.2.2 *Synthetic Data Augmentation*

In cases where the dataset is small, synthetic data can be generated by applying transformations such as:

- **Noise injection:** Draw samples from existing data points and adding random noise to them. This technique is of particular interest if the sensor uncertainty is known, for example from datasheets. Using the specification of the *sensitivity* of the sensors, synthetic datasets can be created.
- **SMOTE (Synthetic Minority Over-sampling Technique):** This heuristic generates synthetic samples by interpolating between existing data points of the minority class in classification tasks [6].
- **Domain Randomization:** Domain randomization is a technique used to generate a wide range of synthetic data with randomized environmental parameters to help models generalize better when applied to real-world scenarios. It is commonly used in vision and autonomous systems. By randomly varying factors such as lighting, object colors, textures, and camera positions in simulated environments, models trained with this augmented data are better equipped to handle variability when deployed in the real world. This method is particularly useful when generating large amounts of synthetic data for domains where real-world data is scarce or expensive to collect.
- **Generative Models:** Advanced techniques, such as Generative Adversarial Networks (GANs) [12], can be employed to generate realistic synthetic data. GANs consist of two neural networks—a generator and a discriminator—that are trained together to produce data indistinguishable from the real data. This approach has been widely used for creating synthetic images, but it can also be applied to other types of data.
- **Geometric Transformations:** In the context of image data, simple transformations such as rotation, scaling, flipping, and translation can be applied to generate new instances. These transformations do not alter the underlying semantics of the

data but introduce variability that can help models generalize. This technique is commonly used in self-supervised learning methods [12] to generate positive pairs of samples.

7.3 K-Fold Cross Validation

K-fold cross-validation is one of the most widely used resampling techniques in machine learning for evaluating the performance of models. The fundamental idea is to partition a dataset into k subsets, called 'folds'. A model is trained k times, each time using $k - 1$ folds for training and the remaining fold for validation. This process provides a robust estimation of model performance, especially when data is limited.

7.3.1 Motivation for K-Fold Cross Validation

The motivation for K-fold cross-validation stems from the need for an unbiased and stable model evaluation. By dividing the dataset into multiple folds and averaging the performance across all folds, the technique helps mitigate overfitting and provides a more reliable estimate of the model's ability to generalize to unseen data.

The key benefits include:

- **Reduced variance:** Instead of depending on a single train-test split, K-fold cross-validation evaluates the model across multiple splits, thus reducing the sensitivity of the performance estimate to any particular partition of the data. This helps provide a more stable and robust performance measure.
- **Efficient use of data:** Unlike a simple train-test split, K-fold cross-validation ensures that each data point is used for both training and testing. Over the course of the k iterations, every observation is used exactly once for validation and k-1 times for training, ensuring maximal use of the available dataset.
- **Statistical evaluation:** Rather than generating a single performance metric (e.g., accuracy or loss), K-fold cross-validation produces k test results, one for each fold. By computing statistics such as the mean and standard deviation across these results, a more comprehensive understanding of the model's generalization capability can be achieved. This statistical approach offers insight into the consistency and variability of the model's performance across different data subsets.

7.3.2 Standard K-Fold Cross Validation

Standard k-fold cross validation is a widely used technique to estimate how well a model generalizes to unseen data. Instead of evaluating a model on a single train–

test split, the dataset is divided into k disjoint subsets, called *folds*. Each fold is used once as a test set, while the remaining $k - 1$ folds are used for training.

This procedure reduces the dependence of the evaluation on a particular data split and provides a more reliable estimate of model performance.

Let D denote the full dataset, which is partitioned into k approximately equal-sized folds:

$$D = \{D_1, D_2, \ldots, D_k\}.$$

The cross-validation procedure is defined as follows:

1. Randomly split the dataset D into k folds $\{D_1, D_2, \ldots, D_k\}$.
2. For each fold $i = 1, \ldots, k$:
   - Use D_i as the test set.
   - Use the remaining data $D \setminus D_i$ as the training set.
   - Train the model on the training set.
   - Evaluate the model on the test set and record the performance metric M_i.
3. Aggregate the performance metrics obtained from all k folds.

   The final performance estimate M is computed as the average over all folds:

$$M = \frac{1}{k} \sum_{i=1}^{k} M_i,$$

where M_i denotes the performance metric (e.g., accuracy, precision, recall, or mean squared error) obtained on the i-th test fold.

Standard k-fold cross validation provides a good trade-off between computational efficiency and reliable performance estimation. Typical values for k are 5 or 10, depending on dataset size and computational cost.

Several variants of k-fold cross validation have been developed to address specific data characteristics or computational constraints. The most common variants are introduced below.

7.3.3 Leave-One-Out Cross Validation (LOO-CV)

Leave-One-Out Cross Validation is a special case of k-fold cross validation where $k = N$, with N being the total number of data points. Each individual sample is used once as a test set, while the remaining $N - 1$ samples form the training set.

This method is particularly useful for very small datasets, where retaining as much training data as possible is critical.

Pseudo code:

- For each data point $i = 1, \ldots, N$:

  - Use sample i as the test set.
  - Use all remaining samples as the training set.
  - Train the model and evaluate its performance.

- Average the performance over all N runs.

Advantages: Nearly unbiased estimate of the generalization error.

Disadvantages: Computationally expensive for large datasets, as the model is trained N times.

7.3.4 *Stratified K-Fold Cross Validation*

Stratified k-fold cross validation ensures that each fold preserves the class distribution of the original dataset. This is especially important for classification tasks with imbalanced class labels.

An illustration of stratified folding is shown in Fig. 7.2b.

Pseudo code:

- Partition the dataset into k folds while preserving class proportions.
- For each fold $i = 1, \ldots, k$:
  - Use fold D_i as the test set.
  - Use all remaining folds as the training set.
  - Train and evaluate the model.
- Compute the average performance across all folds.

Advantages: Produces more reliable results for imbalanced datasets.

Disadvantages: Slightly more complex to implement than standard k-fold cross validation.

7.3.5 *Leave-P-Out Cross Validation*

Leave-P-Out Cross Validation generalizes LOO-CV by leaving out P samples at each iteration. The model is trained on the remaining $N - P$ samples and tested on the excluded subset. This process is repeated for all possible combinations of P samples.

Pseudo code:

- For each combination of P samples:
  - Use the selected P samples as the test set.
  - Use the remaining samples as the training set.
  - Train and evaluate the model.

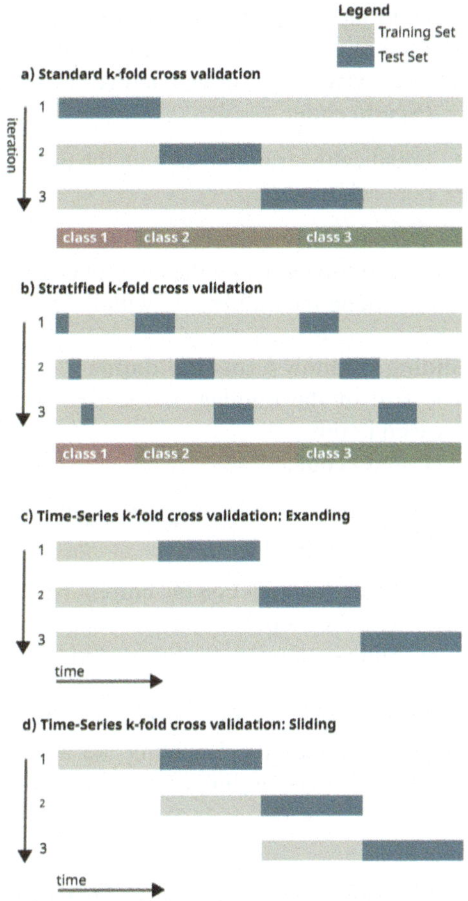

Fig. 7.2 Four cross-validation strategies: (**a**) standard k-fold cross-validation over three iterations; (**b**) stratified k-fold, useful for datasets with class imbalance; (**c**) expanding-window cross-validation for time-series data; and (**d**) sliding-window cross-validation, maintaining a fixed window size while progressing through time

- Average the performance over all combinations.

  Advantages: Very thorough evaluation of model performance.
  Disadvantages: Computationally infeasible for moderate to large P or datasets.

7.3.6 Cross Validation for Time-Series Data

For time-series data, preserving temporal order is essential. Random shuffling, as used in standard k-fold cross validation, may introduce information leakage. Instead, training always uses past data, while testing is performed on future observations.

Two commonly used strategies are illustrated in Fig. 7.2c and d.

Expanding Window Cross Validation
The training set grows over time as more historical data becomes available, while testing is always performed on the immediately following time interval.

Pseudo code:

- Initialize the training set with the earliest observations.
- For each time step:
  - Train the model on all available past data.
  - Test the model on the next time window.
  - Extend the training set with the newly observed data.

Sliding Window Cross Validation
The size of the training and testing sets remains fixed, but the window moves forward in time.

Pseudo code:

- Define fixed-size training and testing windows.
- For each time step:
  - Train the model on the current training window.
  - Test the model on the subsequent testing window.
  - Slide both windows forward in time.

Each cross-validation variant addresses different data characteristics and computational trade-offs. Selecting the appropriate method is essential for obtaining reliable and meaningful performance estimates.

7.4 Practical Exercises: Data Organization

The following exercises are designed to help students gain hands-on experience with key data organization techniques, focusing on data splitting, augmentation, and validation strategies in machine learning workflows.

7.4.1 Exercise 1: Simple Train-Test Split

Objective: Implement a basic train-test split on a given dataset and evaluate the model's performance.

Given is a dataset. Test the following processes:

- Loading the public dataset.
- Split the dataset into training and test sets using a $80-20$ (train-test) ratio.
- Train a simple classification model (e.g., Logistic Regression) on the training set.

- Evaluate the performance on the test set using metrics like accuracy, precision, recall, or F1-score.

  Report the results. Answer the following questions:

- Does the model generalize well to the test dataset?
- Which information is necessary to answer the above question?
- Does the model overfit to the training data?

7.4.2 Exercise 2: Implementing K-Fold Cross-Validation

Objective: Test strategies to avoid overfitting using k-fold cross validation.
Given is a dataset. Test the following processes:

- Loading the public dataset.
- Implement K-fold cross-validation (with $k = 5$).
- Train a simple classification model (e.g., Logistic Regression) on the training set and evaluate if k-times.
- Compare the performance of the model when using different values of k (e.g., $k = 3, k = 10$).
- Plot the training and validation accuracy for each fold.

  Report the results. Answer the following questions:

- Which variant of k-fold cross validation was tested?
- How can one optimize the parameter k?
- What is the optimal value for k for the given model and dataset?

7.4.3 Exercise 3: Stratified K-Fold Cross-Validation

Objective: Test stratified K-fold cross-validation on imbalanced datasets.
Given is a dataset. Test the following processes:

- Loading the public dataset.
- Determine the number of classes and samples per class.
- Visualize the findings.
- Define a conceptual strategy to create balanced training and test datasets.
- Implement your strategy using the pre-implemented stratified K-fold cross-validation method.
- Compare the performance of the model when using different values of k (e.g., $k = 3, k = 10$).
- Plot the training and validation accuracy for each fold.
- Compare the results with regular K-fold cross-validation and discuss the impact on performance metrics such as F1-score.

Fig. 7.3 For the latest online version of the notebook, scan the QR code or visit: https://github.com/mul-cps/DMBook/tree/main/ex_data_organization

Report the results. Answer the following questions:

- What is an imbalanced dataset?
- How could an imaginary example of an imbalanced dataset for classification look like?
- What is the difference to an imbalanced dataset for regression?
- What other strategies do you know to create balanced datasets?

7.5 Jupyter NB Exercise—Data Organization

In this exercise, you will study fundamental techniques for organizing data in machine learning workflows. The focus lies on preparing datasets, splitting data into training and testing subsets, and applying validation strategies to support reliable model evaluation. Through hands-on tasks, you will compare simple train–test splits with more advanced methods such as k-fold and stratified k-fold cross-validation (Fig. 7.3).

Chapter 8
Selected Machine Learning Applications

In this chapter, we present selected applications of machine learning, focusing on real-world scenarios where machine learning techniques are used to derive meaningful insights from data. We explore how machine learning can be applied to solve challenges, particularly when data availability is limited.

8.1 Learning from Very Little Data

In many real-world machine learning applications, the availability of sufficient data for training, validating, and testing models poses a significant challenge. This chapter highlights an application where machine learning is applied to predict the floor level of plant species in green facades of two buildings in Vienna, Austria, using limited datasets. The results presented here are part of a larger study on technology-critical elements (TCEs), with portions previously published in [33].

The research takes a holistic approach to studying TCE distribution in urban ecosystems, including water samples from nearby rivers. However, this section focuses exclusively on modeling TCE accumulation in plant species using machine learning.

We demonstrate that leveraging data augmentation techniques—specifically those based on sensor uncertainty-dependent sampling—can improve prediction accuracy by more than 30% without diluting correlations between input features, which is critical for maintaining the interpretability of the model.

E. Rueckert, *Data Modeling*, https://doi.org/10.1007/978-981-95-7592-3_8

8.1.1 Real-World Problem and Research Question Addressed using Machine Learning

The central research question of this study is: '*How effectively can different plant species in green facades absorb technology-critical elements (TCEs) from urban air?*'

This question is highly relevant in the context of growing environmental concerns surrounding the increasing use of TCEs in modern technologies. These elements, which include lithium (Li), beryllium (Be), gallium (Ga), and germanium (Ge), are critical to high-tech industries but pose significant sustainability and environmental risks due to their scarcity and the ecological impacts of their extraction and distribution.

Understanding how TCEs accumulate in plants is crucial for ecological risk assessment, urban sustainability planning, and efforts to mitigate TCE emissions. Thus, the ability to predict the absorption of these elements by plants is a key step in assessing how urban green infrastructure can mitigate pollution.

8.1.2 Dataset: Sensors and Measurement Data

The dataset used in this study comprises 291 samples of TCE distributions in plant species collected from green facades of two buildings in Vienna. The buildings, referred to as *MA31* and *MA48*, differ in terms of plant species composition and floor levels. As a result, the data were modeled separately for each building, resulting in two datasets:

- **MA31**: 100 samples.
- **MA48**: 191 samples.

Each sample includes 54 features, which are divided into process information and sensor measurements:

8.1.2.1 Process/Descriptive Data (6 Features)

The following features describe the conditions under which the samples were collected:

- **Sample ID**: A unique identifier for each sample (not used in the modeling).
- **Sampling Date**: The date of sample collection, ranging from November 2020 to September 2021.
- **Building**: The building from which the sample was collected, either **MA31** or **MA48**.
- **Level**: The floor level or height from which the plant sample was taken.
- **Plant Species**: The species of the plant from which the sample was obtained.

- **Washed/Unwashed (w/uw)**: Whether the sample was washed before spectrometer analysis to remove surface contaminants.

8.1.2.2 Sensor/Measurement Data (48 Features)

These features represent the measured concentrations of 48 chemical elements, reported as mass fractions in ng/g (nanograms per gram). These elements are crucial for understanding TCE distribution in plants and include:

Elements Li [ng/g], Be [ng/g], Na [ng/g], Mg [ng/g], Al [ng/g], Ca [ng/g], V [ng/g], Cr [ng/g], Mn [ng/g], Fe [ng/g], Co [ng/g], Ni [ng/g], Cu [ng/g], Zn [ng/g], Ga [ng/g], Ge [ng/g], As [ng/g], Se [ng/g], Rb [ng/g], Sr [ng/g], Y [ng/g], Nb [ng/g], Mo [ng/g], Ag [ng/g], Cd [ng/g], Sb [ng/g], Te [ng/g], Ba [ng/g], La [ng/g], Ce [ng/g], Pr [ng/g], Nd [ng/g], Sm [ng/g], Eu [ng/g], Gd [ng/g], Tb [ng/g], Dy [ng/g], Ho [ng/g], Er [ng/g], Tm [ng/g], Yb [ng/g], Lu [ng/g], Ta [ng/g], Tl [ng/g], Pb [ng/g], Bi [ng/g], Th [ng/g], and U [ng/g].

The spectrometer-based measurements capture the mass fractions of each element, providing the basis for assessing the concentration of TCEs in plant tissues.

8.1.3 Machine Learning Approach

The machine learning task was framed as a regression problem, where the goal was to predict the floor level of plant samples based on their measured concentrations of technology-critical elements (TCEs). Various machine learning algorithms, including decision trees, Gaussian processes, and neural networks, were applied and compared in terms of their prediction performance.

Given the relatively small size of the dataset, a key challenge is to improve the model's predictive performance without overfitting. To address this, in this example two key strategies were implemented.

8.1.3.1 Representative Test Samples

To ensure robust model evaluation, the model predictions were tested on a set of *previously unseen test samples* during training. These samples were manually selected by domain experts to represent the diversity of the problem space. By carefully choosing representative test samples, we ensured that the evaluation was more reflective of real-world scenarios, increasing the reliability of the results.

Figure 8.1 shows the model's performance on the representative test samples, illustrating the comparison between the true and predicted values for the target feature, *floor level*. The red dots indicate the manually selected representative samples used for testing.

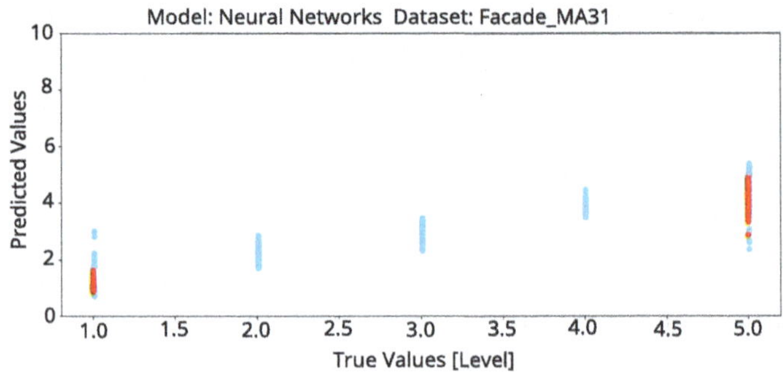

Fig. 8.1 True vs. predicted values for the *floor level* feature in a green facade dataset. Red dots indicate representative samples manually selected for testing

8.1.3.2 Data Augmentation

To further improve model performance, data augmentation techniques were employed. These techniques are based on sensor uncertainty-dependent sampling, which leverages known uncertainties in sensor measurements provided by the spectrometer's specifications to generate synthetic data points. By augmenting the dataset with additional samples that reflect real-world uncertainties, one can effectively increase the amount of training data, enabling the model to better generalize to unseen data while preserving the relationships between features.

The augmentation process maintained the underlying correlations between input features, ensuring the synthetic data did not introduce noise or distort key relationships in the dataset. This was particularly important for maintaining model interpretability and avoiding data distortion.

8.1.4 Results

The primary metric used to evaluate the models was the *mean absolute error (MAE)*, with the target variable, *floor level*, ranging from 1 to 5 (referring to the 1st through 5th floor levels). Figure 8.2 shows the performance of the data augmentation strategy on the MA31 dataset, comparing the results from using large augmented datasets to those from the original smaller dataset.

The study invested the prediction error of the *floor level*. The target values in both datasets were in the range $[1, 5]$, referring to the 1-st till the 5-th floor level. The error is computed as mean squared error.

For the MA31 dataset, the use of sensor uncertainty-dependent sampling improved the model's prediction accuracy by over 30% in terms of MAE compared to baseline models trained solely on the original dataset. Even higher improvements were observed for the larger MA48 dataset, as shown in Fig. 8.3.

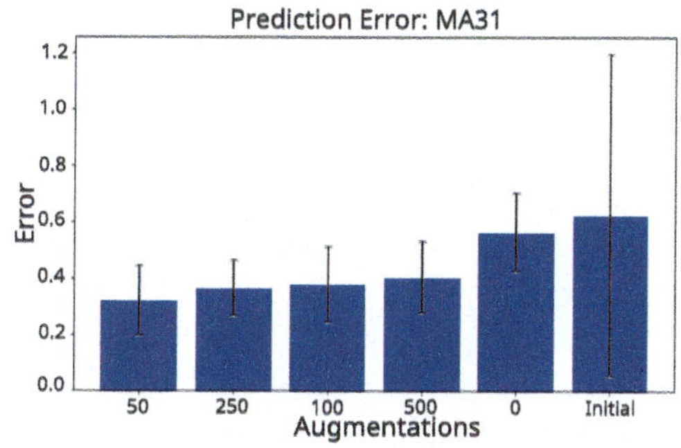

Fig. 8.2 Performance of the data augmentation strategy on the MA31 dataset. The results compare models trained on augmented datasets with those trained on the original small dataset. The difference between 0 and *initial* reflects the effect of hyper-parameter optimization

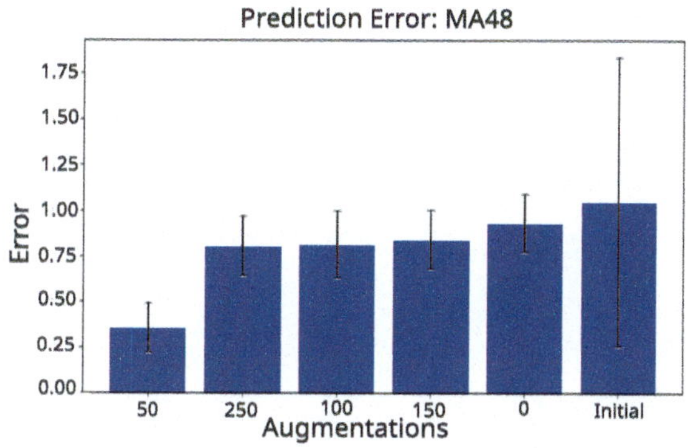

Fig. 8.3 Results of data augmentation strategies for the MA48 dataset

Furthermore, the correlations between the input features (i.e., the measured TCE concentrations) and the target (floor level) were preserved, ensuring that the augmented data did not introduce noise or distort the underlying relationships in the data.

8.1.5 Conclusion

This case study demonstrates the effectiveness of applying machine learning to real-world environmental datasets, even when data is limited. By employing data augmentation techniques, we were able to enhance the predictive performance of the model without sacrificing interpretability. The results provide valuable insights into how plant species in urban green infrastructure can absorb TCEs, with potential implications for urban planning and sustainability initiatives.

Machine learning, combined with careful data preprocessing and augmentation, offers a powerful tool for analyzing complex environmental data and addressing ecological challenges such as the distribution of technology-critical elements in urban ecosystems.

Appendix A
Notation

We follow the *International Conference on Learning Representations* (ICLR) attempt to a standardized notation based on the textbook *Deep Learning* [12].

A.1 Numbers and Arrays

a	A scalar (integer or real)
$\boldsymbol{a}$	A vector
$\boldsymbol{A}$	A matrix
A	A tensor
$\boldsymbol{I}_n$	Identity matrix with n rows and n columns
$\boldsymbol{I}$	Identity matrix with dimensionality implied by context
$\boldsymbol{e}^{(i)}$	Standard basis vector $[0, \ldots, 0, 1, 0, \ldots, 0]$ with a 1 at position i
$\text{diag}(\boldsymbol{a})$	A square, diagonal matrix with diagonal entries given by $\boldsymbol{a}$
a	A scalar random variable
$\mathbf{a}$	A vector-valued random variable
$\mathbf{A}$	A matrix-valued random variable

A.2 Sets and Graphs

$\mathbb{A}$	A set
$\mathbb{R}$	The set of real numbers
$\{0, 1\}$	The set containing 0 and 1
$\{0, 1, \ldots, n\}$	The set of all integers between 0 and n
$[a, b]$	The real interval including a and b
$(a, b]$	The real interval excluding a but including b

E. Rueckert, *Data Modeling*, https://doi.org/10.1007/978-981-95-7592-3

$\mathbb{A}\backslash\mathbb{B}$	Set subtraction, i.e., the set containing the elements of $\mathbb{A}$ that are not in $\mathbb{B}$
$\mathcal{G}$	A graph
$Pa_{\mathcal{G}}(\mathrm{x}_i)$	The parents of x_i in $\mathcal{G}$

A.3 Indexing

a_i	Element i of vector $\boldsymbol{a}$, with indexing starting at 1
a_{-i}	All elements of vector $\boldsymbol{a}$ except for element i
$A_{i,j}$	Element i, j of matrix $\boldsymbol{A}$
$\boldsymbol{A}_{i,:}$	Row i of matrix $\boldsymbol{A}$
$\boldsymbol{A}_{:,i}$	Column i of matrix $\boldsymbol{A}$
$\mathsf{A}_{i,j,k}$	Element (i, j, k) of a 3-D tensor A
$\mathsf{A}_{:,:,i}$	2-D slice of a 3-D tensor
a_i	Element i of the random vector $\mathbf{a}$

A.4 Calculus

$\frac{dy}{dx}$	Derivative of y with respect to x
$\frac{\partial y}{\partial x}$	Partial derivative of y with respect to x
$\nabla_{\boldsymbol{x}} y$	Gradient of y with respect to $\boldsymbol{x}$
$\nabla_{\boldsymbol{X}} y$	Matrix derivatives of y with respect to $\boldsymbol{X}$
$\nabla_{\mathsf{X}} y$	Tensor containing derivatives of y with respect to X
$\frac{\partial f}{\partial \boldsymbol{x}}$	Jacobian matrix $\boldsymbol{J} \in \mathbb{R}^{m \times n}$ of $f : \mathbb{R}^n \to \mathbb{R}^m$
$\nabla_{\boldsymbol{x}}^2 f(\boldsymbol{x})$ or $\boldsymbol{H}(f)(\boldsymbol{x})$	The Hessian matrix of f at input point $\boldsymbol{x}$
$\int f(\boldsymbol{x}) d\boldsymbol{x}$	Definite integral over the entire domain of $\boldsymbol{x}$
$\int_{\mathbb{S}} f(\boldsymbol{x}) d\boldsymbol{x}$	Definite integral with respect to $\boldsymbol{x}$ over the set $\mathbb{S}$

A.5 Probability and Information Theory

$P(\mathrm{a})$	A probability distribution over a discrete variable
$p(\mathrm{a})$	A probability distribution over a continuous variable, or over a variable whose type has not been specified
$\mathrm{a} \sim P$	Random variable a has distribution P

$\mathbb{e}_{\mathrm{x}\sim P}[f(x)]$ or $\mathbb{e} f(x)$	Expectation of $f(x)$ with respect to $P(\mathrm{x})$
$\mathrm{Var}(f(x))$	Variance of $f(x)$ under $P(\mathrm{x})$
Σ or $\mathrm{Cov}(f(x), g(x))$	Covariance of $f(x)$ and $g(x)$ under $P(\mathrm{x})$
$H(\mathrm{x})$	Shannon entropy of the random variable x
$D_{\mathrm{KL}}(P \| Q)$	Kullback-Leibler divergence of P and Q
$\mathcal{N}(\boldsymbol{x}; \boldsymbol{\mu}, \boldsymbol{\Sigma})$	Gaussian distribution over $\boldsymbol{x}$ with mean $\boldsymbol{\mu}$ and covariance $\boldsymbol{\Sigma}$

A.6 Functions

$f : \mathbb{A} \to \mathbb{B}$	The function f with domain $\mathbb{A}$ and range $\mathbb{B}$
$f \circ g$	Composition of the functions f and g
$f(\boldsymbol{x}; \boldsymbol{\theta})$	A function of $\boldsymbol{x}$ parametrized by $\boldsymbol{\theta}$. (Sometimes we write $f(\boldsymbol{x})$ and omit the argument $\boldsymbol{\theta}$ to lighten notation)
$\log x$	Natural logarithm of x
$\sigma(x)$	Logistic sigmoid, $\dfrac{1}{1+\exp(-x)}$
$\zeta(x)$	Softplus, $\log(1+\exp(x))$
$\|\|\boldsymbol{x}\|\|_p$	L^p norm of $\boldsymbol{x}$
$\|\|\boldsymbol{x}\|\|$	L^2 norm of $\boldsymbol{x}$
x^+	Positive part of x, i.e., $\max(0, x)$
$\mathbf{1}_{\mathrm{condition}}$	is 1 if the condition is true, 0 otherwise

A.7 The Greek Alphabet

Character	Name	Character	Name
α	alpha *AL-fuh*	ν	nu *NEW*
β	beta *BAY-tuh*	ξ, Ξ	xi *KSIGH*
γ, Γ	gamma *GAM-muh*	o	omicron *OM-uh-CRON*
δ, Δ	delta *DEL-tuh*	π, Π	pi *PIE*
ϵ	epsilon *EP-suh-lon*	ρ	rho *ROW*
ζ	zeta *ZAY-tuh*	σ, Σ	sigma *SIG-muh*
η	eta *AY-tuh*	τ	tau *TOW (as in cow)*
θ, Θ	theta *THAY-tuh*	υ, Υ	upsilon *OOP-suh-LON*
ι	iota *eye-OH-tuh*	ϕ, Φ	phi *FEE, or FI (as in hi)*
κ	kappa *KAP-uh*	χ	chi *KI (as in hi)*
λ, Λ	lambda *LAM-duh*	ψ, Ψ	psi *SIGH, or PSIGH*
μ	mu *MEW*	ω, Ω	omega *oh-MAY-guh*

Capitals shown are the ones that differ from Roman capitals

Appendix B
Learning Objectives

B.1 Learning Objectives of This Book

This book covers the following learning objectives:

1. **Understanding the Role of Physical Processes in Data Modeling**
   - Describe and explain physical processes.
   - Understand process abstractions and sub-processes.
   - Identify and define assumptions and limitations in modeling.
2. **Understanding the Function of Sensors in Data Collection**
   - Explain key sensing technologies such as:
     - Optical sensors (e.g., cameras, photodiodes)
     - Acoustic sensors (e.g., microphones, ultrasonic sensors)
     - Temperature sensors (e.g., thermocouples, RTDs)
     - Pressure sensors (e.g., piezoelectric, capacitive)
     - Magnetic sensors (e.g., Hall effect)
     - Inertial sensors (e.g., accelerometers, gyroscopes)
   - Identify suitable sensors for typical applications in data modeling.
   - Assess assumptions and limitations of sensor technologies in context.
3. **Transforming Raw Data into Structured Variables**
   - Understand data types, sampling frequencies, memory requirements, and conversions.
   - Predict memory requirements for given sensing configurations.
   - Organize data in object-oriented structures to avoid redundancy.
   - Understand lossless and lossy data compression methods.
   - Gain familiarity with large-scale data storage, distributed systems, and data centers.

E. Rueckert, *Data Modeling*, https://doi.org/10.1007/978-981-95-7592-3

4. **Preprocessing and Analyzing Data**
   - Load and preprocess datasets using data imputation and filtering.
   - Compute data statistics and identify anomalies.
   - Visualize data using histograms, correlation plots, and low-dimensional projections.
   - Apply data augmentation to extend datasets and reflect modeling assumptions.
5. **Modeling Data for Machine Learning Algorithms**
   - Perform train/validation/test splits based on application-specific knowledge.
   - Implement validation techniques such as k-fold cross-validation.
   - Understand data correlations, especially temporal dependencies.
   - Identify modeling pitfalls including overfitting, long-tail distributions, and the No Free Lunch theorem.
6. **Understanding Machine Learning Algorithms and Evaluation Metrics**
   - Explain fundamental machine learning approaches.
   - Understand evaluation metrics and benchmarking practices.
   - Identify and optimize hyperparameters.
   - Design evaluation strategies aligned with modeling goals.

Appendix C
Mathematical Background and Derivations

In this section, we discuss some basic concepts that will be used in the course.

C.1 Two-Measurement Model of the Gummi Bear Free Fall Experiment

Given two measurements, the system can be modeled using:

$$\begin{pmatrix} t_1 \\ t_2 \end{pmatrix} = \begin{pmatrix} 1 & m_1 \\ 1 & m_2 \end{pmatrix} \begin{pmatrix} c_1 \\ c_2 \end{pmatrix}$$

Solving for the parameters:

$$\begin{pmatrix} c_1 \\ c_2 \end{pmatrix} = \begin{pmatrix} 1 & m_1 \\ 1 & m_2 \end{pmatrix}^{-1} \begin{pmatrix} t_1 \\ t_2 \end{pmatrix}$$

The inverse of a 2×2 matrix is:

$$\begin{pmatrix} a & b \\ c & d \end{pmatrix}^{-1} = \frac{1}{ad - bc} \begin{pmatrix} d & -b \\ -c & a \end{pmatrix}$$

Applying this gives:

$$\begin{pmatrix} c_1 \\ c_2 \end{pmatrix} = \frac{1}{m_2 - m_1} \begin{pmatrix} m_2 & -m_1 \\ -1 & 1 \end{pmatrix} \begin{pmatrix} t_1 \\ t_2 \end{pmatrix}$$

E. Rueckert, *Data Modeling*, https://doi.org/10.1007/978-981-95-7592-3

Explicitly:

$$c_1 = \frac{m_2 t_1 - m_1 t_2}{m_2 - m_1}, \quad c_2 = \frac{t_2 - t_1}{m_2 - m_1}$$

C.2 General Least Squares Model Derivation

To model the full experiment with N measurements, we define:

$$\mathbf{t} = \mathbf{X}\mathbf{c}, \quad \text{with} \quad \mathbf{X} = \begin{pmatrix} 1 & m_1 \\ 1 & m_2 \\ \vdots & \vdots \\ 1 & m_N \end{pmatrix}, \quad \mathbf{c} = \begin{pmatrix} c_1 \\ c_2 \end{pmatrix}$$

Define the residual:

$$\mathbf{e} = \mathbf{t} - \mathbf{X}\mathbf{c}$$

Minimize the squared error:

$$E = \mathbf{e}^\top \mathbf{e} = (\mathbf{t} - \mathbf{X}\mathbf{c})^\top (\mathbf{t} - \mathbf{X}\mathbf{c})$$

Expanding:

$$E = \mathbf{t}^\top \mathbf{t} - 2\mathbf{c}^\top \mathbf{X}^\top \mathbf{t} + \mathbf{c}^\top \mathbf{X}^\top \mathbf{X}\mathbf{c}$$

Take the derivative with respect to $\mathbf{c}$ and set to zero:

$$\frac{\partial E}{\partial \mathbf{c}} = -2\mathbf{X}^\top \mathbf{t} + 2\mathbf{X}^\top \mathbf{X}\mathbf{c} = 0$$

Yielding the normal equation:

$$\mathbf{X}^\top \mathbf{X}\mathbf{c} = \mathbf{X}^\top \mathbf{t}$$

Solving:

$$\mathbf{c} = (\mathbf{X}^\top \mathbf{X})^{-1} \mathbf{X}^\top \mathbf{t}$$

This is the classic **least squares solution**, where the model parameters $\mathbf{c}$ are obtained via the **pseudo-inverse** of $\mathbf{X}$.

Appendix D
Tutorials

This chapter introduces simple techniques and implementations in the field of data modeling. It also provides a beginner-friendly introduction to programming, designed to help readers overcome the initial challenges and get started with coding.

D.1 A Simple Introduction to Programming

Entering the world of programming can feel daunting at first. This tutorial offers a step-by-step introduction with intuitive examples and visual aids to help you grasp fundamental concepts.

D.1.1 What Is a Program?

A program is simply a sequence of instructions that a computer follows to solve a problem or perform a task. More formally, a program is an implementation of an algorithm—a set of rules or procedures for converting inputs into desired outputs. However, in its simplest form, a program may not involve inputs or outputs and may consist only of a few basic commands.

As an analogy, think of a light switch that controls a light in a room. A basic program could follow two steps:

As an example, consider a light switch that turns on or off the light in a room. The commands used in this simple program are

- **Checks**: Check if the switch is on or off (condition).
- **State Assignments**: Turn the light on or off (action).

These two compontens are elementary principles in programming. They are refered as conditional statements (checks) and variable assignments (state assignments).

E. Rueckert, *Data Modeling*, https://doi.org/10.1007/978-981-95-7592-3

D.1.2 Conditional Statements

Conditional statements allow programs to make decisions based on conditions. This is often done using `'if-then-else'` constructs. Such statements allow you to control what happens depending on whether a condition is true or false.

Conditional statement consosts of three components:

- **Query (Condition)**: This is the logic test that follows the `'if'` statement. For example, `'if(light > 0)'` checks whether the light is currently on.
- **True Condition**: If the query is true, the following commands will be executed. For example, if the light is on, the program may turn it off.
- **False Condition (Optional)**: If the query is false, commands in the `'else'` block are executed instead.

Conditional statements are central to most programs and control the flow of decisions.

D.1.3 Variable Assignments

In programming, variables are used to store and manipulate data. You can think of a variable as a named box that holds a value. The value stored in the variable can change as the program runs.

Variables have two important concepts:

- **Variable Definition**: This is where the variable is introduced, allocating space in memory. For example, `'int lightstate'` defines an integer variable that stores the current state of the light as natural number. The term 'int' denotes the data type (see Sect. 4.2 for other types and definitions).
- **Variable Assignment**: Assigning a value to a variable happens with the '=' symbol. For example, `'lightstate = 1'` sets the light state to 'on'.

Variables are crucial because they allow programs to store data and use it later. In the light switch example, a variable can store whether the light is currently on or off, and conditional statements can then operate based on the variable's value.

D.1.4 Loops

One of the most important constructs in programming is the loop. Loops allow you to repeat sections of code multiple times without having to manually rewrite the instructions. This is especially useful when performing repetitive tasks or processing large datasets.

In most programs, loops are essential for coordinating or directing the flow of execution, ensuring that tasks are performed efficiently. A common use case for loops is keeping a program running continuously, such as in the case of an endless loop for systems that need to stay operational.

Common loops or usages are:

- **Infinite Loops**: An infinite loop is a loop that continues indefinitely because its condition is always true. This is often used in systems where the program must keep running until manually stopped, such as in servers, game loops, or real-time monitoring systems. For example, the statement `'while(1)'` is always true and whatever is inside this loop statement will be executed without termination.
- **For Loops**: A for loop is used when you need to iterate over a collection of items, such as an array, or perform a block of code a set number of times. This type of loop is particularly useful when the number of iterations is known in advance.

When designing loops, always ensure that the loop condition is well-defined to prevent unwanted infinite loops, and use control mechanisms like break or continue statements to handle exceptions or early exits. Loops should be optimized for performance when dealing with large datasets to avoid slowing down your program unnecessarily.

With loops, you gain the ability to automate repetitive tasks, which is essential for building efficient and scalable programs. By mastering loops, you can handle a wide range of scenarios from simple iterations to complex, real-time systems.

D.1.5 Visual, Block-Based Programming

One of the most intuitive ways to learn programming is through visual, block-based environments like *Scratch* [23]. Instead of writing code, you combine blocks representing commands, making programming easier to understand for beginners.

For instance, in *Scratch*, a simple program could be designed to check whether the 'a' key is pressed to turn the light on, or the 'b' key to turn it off.

D.1.6 Playground LEGO EV3

Learning to program becomes even more engaging when paired with interactive hardware. LEGO Mindstorms EV3 is an excellent platform for beginners to apply programming concepts using robots and physical devices.

In this example, we'll build a stopwatch using the EV3, with two buttons: one to start the timer and another to stop it. This practical task teaches core programming concepts such as input handling, variable assignment, and time measurement.

Appendix E
Fact Sheet

This chapter serves as a concise summary of the most important concepts, definitions, and key takeaways from this book. It is designed to help reinforce your understanding and may prove particularly useful when preparing for the final exam.

The content is organized according to the six main components of the book, with each section highlighting the essential facts and ideas covered.

E.1 Real Processes

E.1.1 Terminology Granule and Granularity

- Explain term Granule, Provide Examples
- Explain term Granularity, Provide Examples

E.1.2 Digital Model, Digital Shadow, Digital Twin

- Explain the term Digital Model, Provide Examples, Discuss the Information flow
- Explain the term Digital Shadow, Provide Examples, Discuss the Information flow
- Explain the term Digital Twin, Provide Examples, Discuss the Information flow

E.2 Sensors

E.2.1 Sensor Categorizations

- Which categorizations were discussed, Provide examples

E. Rueckert, *Data Modeling*, https://doi.org/10.1007/978-981-95-7592-3

- What is a stimulus?
- What is a process variable and how does it relate to measurements?

E.2.2 Terminology of Sensors

- Explain term Accuracy, Provide Examples
- Explain term Precision, Provide Examples
- Explain term Sensitivity, Provide Examples
- Explain term Range, Provide Examples
- Compared to humans, how do technologies perform?

E.2.3 Sensor Transfer Function

- Describe the transfer function mathematically
- Which terms are involved?
- Provide examples.
- How does sensor noise influence the model predictions?
- What is the error propagation law? Explain it in your own words.
- Provide examples, of utilizing knowledge of the sensor noise to determine lower and upper bounds on the model prediction errors.

E.3 Data Storage

E.3.1 Terminology of Objects, Attributes and Relations

- Explain Attributes, Provide examples
- Explain Objects, Provide examples
- What are data relations, Provide examples
- Explain 1-to-1, 1-to-many, and many-to-many, provide examples

E.3.2 Entity-Relationship Diagram (ER-Diagram)

- Explain the concept of ER-Diagrams, Provide an example
- Explain 1-to-1, 1-to-many, and many-to-many, provide examples
- Given an ER-diagram, provide an intuitive explanation of the relations

E.3.3 Other Relationship Visualizations

- Explain Universal Modeling Language (UML) Diagrams, what is the difference to ERDs?
- Explain Graph Databases, what are the nodes and links?
- Given an ER-diagram, provide an intuitive explanation of the relations

E.3.4 Data Types and Memory Requirements

- Explain primitive data types, provide examples
- Explain non-primitive data types, provide examples
- Compute storage requirements given a practical example

E.3.5 Compression Methods and Data Centers

- Which types of compression methods were discussed?
- Explain Lossless Data Compression, provide examples
- Explain Lossy Data Compression, provide examples
- Why are data centers so important nowadays?

E.3.6 Database Development

- What are databases?
- Which types of databases were discussed?
- How can ER-Diagrams be used to develop databases, Provide examples.
- What is a primary key?
- Explain auto-increment primary keys
- What are general design rules for databases?

E.3.7 Document-Based Data Storage

- Which types of document-based data storage systems were discussed?
- What are the benefits and the drawbacks of spreadsheets?
- How can spreadsheets be used with databases (import, export)?
- What is a CSV file?
- What are the benefits and the drawbacks of CSV files for data storage?

E.4 Information Theory

E.4.1 Terminology of Bits, Bytes, and Possibilities

- Explain bits and bytes, Provide examples
- Explain the term 'possibilities' with respect to the number of bits used, Provide examples
- Given the required number of possibilities, the need bits are computed as $n = \log_2(x)$

E.4.2 Terminology of Surprise

- Explain the term 'surprise' with respect to the used number of bits
- How does the likelihood of an event relate to the concept of surprise?
- The 'surprise' in bits is computed via $n = \log_2(x)$

E.4.3 Shannon Information

- Explain the term 'Shannon Information'
- How does the term relate to the concept of surprise?
- The Shannon Information is given by $I(X) = \log_2(1/p(X)) = -\log_2(p(X))$
- What happens if the possibilities of events are not equally likely, provide examples
- Explain the term 'Shannon Entropy'
- The Shannon Entropy is given by $H(X) = -\sum_{x \in X} p(x)\log_2(p(x))$

E.4.4 Entropy-Based Compression

- How can the Shannon Entropy be used to improve the data compression, provide an example
- The 'Average-Bits-Per-Symbol (ABS) measure' is given by $ABS = \sum_{x \in X} p(x)L_x$, where L_x denotes the length in bits of x
- What is a 'Fixed-length encoding', provide an example
- What is a 'Variable-length encoding', provide an example
- How does the 'Shannon Entropy' relate to an optimal encoding?

E.4.5 Analog to Digital Conversion

- What is the Nyquist Rate, provide an example
- Explain the Nyquist-Shannon Theorem in own words
- The 'Average-Bits-Per-Symbol (ABS) measure' is given by $ABS = \sum_{x \in X} p(x)L_x$, where L_x denotes the length in bits of x
- What is a 'Fixed-length encoding', provide an example
- What is a 'Variable-length encoding', provide an example
- How does the 'Shannon Entropy' relate to an optimal encoding?

E.5 Analysis

E.5.1 Basic Statistics

- Explain the terms 'Mean', 'Median', 'Mode', 'Variance', 'Standard Deviation', and 'Range'
- Given an equation, explain the terms
- Provide examples

E.5.2 Data Visualization

- What is the issue with plotting statistics?
- How can raw data help interpreting the statistical results?
- Which visualization strategies were discussed (Histograms, Correlation or confusion plots, bar plots, time-series plots)?

E.5.3 Data Normalization and Standardization

- Why is it important to reshape (e.g. normalize) the data (reusing experts knowledge e.g. parameter settings of models)?
- Explain 'normalization' (known as min-max scaling), Provide examples, $x' = (x - xmin)/(xmax - xmin)$
- Explain 'standardization' (known as Z-score normalization), Provide examples, $x' = (x - \mu)/\sigma$
- Which visualization strategies were discussed (Histograms, Correlation or confusion plots)?
- Why is it sometimes useful to transform the data, e.g. using the 'box-cox transformation' (reason Gaussian data assumption)?
- Explain the Z-score outlier detection approach?

E.6 Data Organization

E.6.1 Terminology of Datasets

- Explain the term 'training set', Provide examples
- Explain the term 'validation set', Provide examples
- Explain the term 'test set', Provide examples
- What is the underlying assumption when splitting datasets (answer: all splits are from the same distribution, or all sets equally well cover the whole space)?

E.6.2 Cross Validation

- Explain the approach 'k-fold cross validation', Provide examples
- When is k-fold CV useful (answer: small datasets, avoiding overfitting, to compute statistics)?
- Explain the approach 'stratified k-fold cross validation', Provide examples
- Explain the 'expanding window' approach for time-series k-fold cross validation, Provide examples
- Explain the 'sliding window' approach for time-series k-fold cross validation, Provide examples

E.6.3 Data Augmentation

- Explain the data augmentation term, Provide examples
- What are motivations for data augmentation, Provide examples

E.7 Machine Learning

E.7.1 Examples

- Given an example, define the six components of data modeling
- Which examples were discussed in the course (steel production data and chemistry data of plants on facades)?
- Which questions are important when starting a new data modeling problem (process, sensors, data, outlier, organization, methods, evaluation)?

References

1. Barnes, J. (2000). *Aristotle: A very short introduction* (Vol. 32). Oxford Paperbacks.
2. Batty, M. (2018). Digital twins. *Environment and Planning B: Urban Analytics and City Science, 45*(5), 817–820.
3. Bishop, C. M. (2006). *Pattern recognition and machine learning*. Springer.
4. Bosch Sensortec: Bosch sensortec bma400 accelerometer (2024). https://raw.githubusercontent.com/mul-cps/DMBook/main/Datasheets/bst-bma400-ds000.pdf, last visited Dec. 2025.
5. Capozziello, S., & De Laurentis, M. (2011). Extended theories of gravity. *Physics Reports, 509*(4–5), 167–321.
6. Chawla, N. V., Bowyer, K. W., Hall, L. O., & Kegelmeyer, W. P. (2002). Smote: Synthetic minority over-sampling technique. *Journal of Artificial Intelligence Research, 16*, 321–357.
7. Connolly, T. M., & Begg, C. E. (2005). *Database systems: a practical approach to design, implementation, and management*. Pearson Education.
8. Drake, S. (1973). Galileo's discovery of the law of free fall. *Scientific American, 228*(5), 84–93.
9. Fraden, J., & King, J. (2004). *Handbook of modern sensors: physics, designs, and applications* (Vol. 3). Springer.
10. Freedman, D. A. (2009). *Statistical Models: Theory and Practice*. Cambridge University Press.
11. Garling, D. J. (2007). *Inequalities: A Journey into Linear Analysis*. Cambridge University Press.
12. Goodfellow, I., Bengio, Y., & Courville, A. (2016). *Deep Learning*. MIT Press. http://www.deeplearningbook.org.
13. Goulet, J. A. (2020). Probabilistic machine learning for civil engineers. MIT Press.
14. Halliday, D., Resnick, R., & Walker, J. (2013). *Fundamentals of physics*. John Wiley & Sons.
15. Honeywell International Inc. (2024). Honeywell air flow sensor AWM700 series. https://raw.githubusercontent.com/mul-cps/DMBook/main/Datasheets/sps-siot-airflow-sensors-awm700-series-datasheet-32301627-b-en-ciid-50061.pdf, last visited Dec. 2025.
16. Honeywell International Inc. (2024). Honeywell fss series force sensor. https://raw.githubusercontent.com/mul-cps/DMBook/main/Datasheets/ForceSensorHoneywellFSSeries.pdf, last visited Dec. 2025.
17. Infineon Technologies. (2024). Infineon BGT24MTR11 24GHz radar sensor. https://raw.githubusercontent.com/mul-cps/DMBook/main/Datasheets/infineon-bgt24mtr11-datasheet-en.pdf, last visited Dec. 2025.
18. Intel Corporation. (2024). RGB-D camera intel realsense D435i. https://raw.githubusercontent.com/mul-cps/DMBook/main/Datasheets/Intel-RealSense-D400-Series-Datasheet.pdf, last visited Dec. 2025.

E. Rueckert, *Data Modeling*, https://doi.org/10.1007/978-981-95-7592-3

19. Joy-IT GmbH. (2024). Sen-MQ135 gas sensor. https://raw.githubusercontent.com/mul-cps/DMBook/main/Datasheets/SEN-MQ135_Datasheet_2023-09-26.pdf, last visited Dec. 2025.
20. Kravets, A. G., Bolshakov, A. A., & Shcherbakov, M. V. (2020). *Cyber-physical systems: Advances in design & modelling*. Springer.
21. Larose, D. T., & Larose, C. D. (2014). *Discovering knowledge in data: an introduction to data mining*. John Wiley & Sons.
22. Lindberg, D. C. (2010). *The beginnings of Western science: The European scientific tradition in philosophical, religious, and institutional context, prehistory to AD 1450*. University of Chicago Press.
23. MIT Media Lab. (2024). Scratch programming – mit. https://scratch.mit.edu, last visited Oct. 2024.
24. Murphy, K. P. (2012). *Machine learning: a probabilistic perspective*. MIT Press.
25. Pedrycz, W., Chen, S. M., et al. (2015). *Information granularity, big data, and computational intelligence*. Springer.
26. Rueckert, E., Čamernik, J., Peters, J., & Babič, J. (2016). Probabilistic movement models show that postural control precedes and predicts volitional motor control. *Scientific Reports, 6*(1), 28455.
27. Schrenk, M. (2004). Galileo versus aristotle on free falling bodies. *History of Philosophy & Logical Analysis, 7*(1), 81–89. https://doi.org/10.30965/26664275-00701006. Examines historical views on motion and the shift from Aristotelian to Galilean theories.
28. Sensirion AG. (2025). Humidity and temperature sensor sensirion SHT41. https://raw.githubusercontent.com/mul-cps/DMBook/main/Datasheets/HT_DS_Datasheet_SHT4x_5.pdf, last visited Dec. 2025.
29. Serway, R. A., Jewett, J. W., & Peroomian, V. (2000). *Physics for scientists and engineers* (Vol. 2). Saunders College Publishing.
30. STMicroelectronics. (2024). Stmicroelectronics MP23DB01HP. https://raw.githubusercontent.com/mul-cps/DMBook/main/Datasheets/mp23db01hp.pdf, last visited Dec. 2025.
31. Teorey, T. J. (1999). *Database modeling and design*. Morgan Kaufmann.
32. Timón-Reina, S., Rincón, M., & Martínez-Tomás, R. (2021). An overview of graph databases and their applications in the biomedical domain. *Database, 2021*, baab026. https://doi.org/10.1093/database/baab026.
33. Trimmel, S., Spörl, P., Haluza, D., Lashin, N., Meisel, T. C., Pitha, U., Prohaska, T., Puschenreiter, M., Rückert, E., Spangl, B., et al. (2024). Green and blue infrastructure as model system for emissions of technology-critical elements. *Science of The Total Environment, 934*, 173364.
34. Xu, Y., & Goodacre, R. (2018). On splitting training and validation set: a comparative study of cross-validation, bootstrap and systematic sampling for estimating the generalization performance of supervised learning. *Journal of Analysis and Testing, 2*(3), 249–262.

Index

E. Rueckert, *Data Modeling*, https://doi.org/10.1007/978-981-95-7592-3

The manufacturer's authorised representative in the EU is Springer Nature Customer Service Centre GmbH, Europaplatz 3, 69115 Heidelberg, Germany. If you have any concerns regarding our products, please contact ProductSafety@springernature.com

Printed and bound by CPI Group (UK) Ltd, Croydon, CR0 4YY
09/08/2026
02190805-0001